FREE Test Taking Tips DVD Offer

To help us better serve you, we have developed a Test Taking Tips DVD that we would like to give you for FREE. **This DVD covers world-class test taking tips that you can use to be even more successful when you are taking your test.**

All that we ask is that you email us your feedback about your study guide. Please let us know what you thought about it – whether that is good, bad or indifferent.

To get your **FREE Test Taking Tips DVD**, email freedvd@studyguideteam.com with "FREE DVD" in the subject line and the following information in the body of the email:

a. The title of your study guide.

b. Your product rating on a scale of 1-5, with 5 being the highest rating.

c. Your feedback about the study guide. What did you think of it?

d. Your full name and shipping address to send your free DVD.

If you have any questions or concerns, please don't hesitate to contact us at freedvd@studyguideteam.com.

Thanks again!

Middle Level ISEE Test Prep

ISEE Study Guide with Practice Questions for the Independent School Entrance Exam [3rd Edition]

Joshua Rueda

Written and edited by TPB Publishing.

ISBN 13: 9781628457391
ISBN 10: 1628457392

Table of Contents

Quick Overview

As you draw closer to taking your exam, effective preparation becomes more and more important. Thankfully, you have this study guide to help you get ready. Use this guide to help keep your studying on track and refer to it often.

This study guide contains several key sections that will help you be successful on your exam. The guide contains tips for what you should do the night before and the day of the test. Also included are test-taking tips. Knowing the right information is not always enough. Many well-prepared test takers struggle with exams. These tips will help equip you to accurately read, assess, and answer test questions.

A large part of the guide is devoted to showing you what content to expect on the exam and to helping you better understand that content. In this guide are practice test questions so that you can see how well you have grasped the content. Then, answer explanations are provided so that you can understand why you missed certain questions.

Don't try to cram the night before you take your exam. This is not a wise strategy for a few reasons. First, your retention of the information will be low. Your time would be better used by reviewing information you already know rather than trying to learn a lot of new information. Second, you will likely become stressed as you try to gain a large amount of knowledge in a short amount of time. Third, you will be depriving yourself of sleep. So be sure to go to bed at a reasonable time the night before. Being well-rested helps you focus and remain calm.

Be sure to eat a substantial breakfast the morning of the exam. If you are taking the exam in the afternoon, be sure to have a good lunch as well. Being hungry is distracting and can make it difficult to focus. You have hopefully spent lots of time preparing for the exam. Don't let an empty stomach get in the way of success!

When travelling to the testing center, leave earlier than needed. That way, you have a buffer in case you experience any delays. This will help you remain calm and will keep you from missing your appointment time at the testing center.

Be sure to pace yourself during the exam. Don't try to rush through the exam. There is no need to risk performing poorly on the exam just so you can leave the testing center early. Allow yourself to use all of the allotted time if needed.

Remain positive while taking the exam even if you feel like you are performing poorly. Thinking about the content you should have mastered will not help you perform better on the exam.

Once the exam is complete, take some time to relax. Even if you feel that you need to take the exam again, you will be well served by some down time before you begin studying again. It's often easier to convince yourself to study if you know that it will come with a reward!

Test-Taking Strategies

1. Predicting the Answer

When you feel confident in your preparation for a multiple-choice test, try predicting the answer before reading the answer choices. This is especially useful on questions that test objective factual knowledge. By predicting the answer before reading the available choices, you eliminate the possibility that you will be distracted or led astray by an incorrect answer choice. You will feel more confident in your selection if you read the question, predict the answer, and then find your prediction among the answer choices. After using this strategy, be sure to still read all of the answer choices carefully and completely. If you feel unprepared, you should not attempt to predict the answers. This would be a waste of time and an opportunity for your mind to wander in the wrong direction.

2. Reading the Whole Question

Too often, test takers scan a multiple-choice question, recognize a few familiar words, and immediately jump to the answer choices. Test authors are aware of this common impatience, and they will sometimes prey upon it. For instance, a test author might subtly turn the question into a negative, or he or she might redirect the focus of the question right at the end. The only way to avoid falling into these traps is to read the entirety of the question carefully before reading the answer choices.

3. Looking for Wrong Answers

Long and complicated multiple-choice questions can be intimidating. One way to simplify a difficult multiple-choice question is to eliminate all of the answer choices that are clearly wrong. In most sets of answers, there will be at least one selection that can be dismissed right away. If the test is administered on paper, the test taker could draw a line through it to indicate that it may be ignored; otherwise, the test taker will have to perform this operation mentally or on scratch paper. In either case, once the obviously incorrect answers have been eliminated, the remaining choices may be considered. Sometimes identifying the clearly wrong answers will give the test taker some information about the correct answer. For instance, if one of the remaining answer choices is a direct opposite of one of the eliminated answer choices, it may well be the correct answer. The opposite of obviously wrong is obviously right! Of course, this is not always the case. Some answers are obviously incorrect simply because they are irrelevant to the question being asked. Still, identifying and eliminating some incorrect answer choices is a good way to simplify a multiple-choice question.

4. Don't Overanalyze

Anxious test takers often overanalyze questions. When you are nervous, your brain will often run wild, causing you to make associations and discover clues that don't actually exist. If you feel that this may be a problem for you, do whatever you can to slow down during the test. Try taking a deep breath or counting to ten. As you read and consider the question, restrict yourself to the particular words used by the author. Avoid thought tangents about what the author *really* meant, or what he or she was *trying* to say. The only things that matter on a multiple-choice test are the words that are actually in the question. You must avoid reading too much into a multiple-choice question, or supposing that the writer meant something other than what he or she wrote.

5. No Need for Panic

It is wise to learn as many strategies as possible before taking a multiple-choice test, but it is likely that you will come across a few questions for which you simply don't know the answer. In this situation, avoid panicking. Because most multiple-choice tests include dozens of questions, the relative value of a single wrong answer is small. As much as possible, you should compartmentalize each question on a multiple-choice test. In other words, you should not allow your feelings about one question to affect your success on the others. When you find a question that you either don't understand or don't know how to answer, just take a deep breath and do your best. Read the entire question slowly and carefully. Try rephrasing the question a couple of different ways. Then, read all of the answer choices carefully. After eliminating obviously wrong answers, make a selection and move on to the next question.

6. Confusing Answer Choices

When working on a difficult multiple-choice question, there may be a tendency to focus on the answer choices that are the easiest to understand. Many people, whether consciously or not, gravitate to the answer choices that require the least concentration, knowledge, and memory. This is a mistake. When you come across an answer choice that is confusing, you should give it extra attention. A question might be confusing because you do not know the subject matter to which it refers. If this is the case, don't eliminate the answer before you have affirmatively settled on another. When you come across an answer choice of this type, set it aside as you look at the remaining choices. If you can confidently assert that one of the other choices is correct, you can leave the confusing answer aside. Otherwise, you will need to take a moment to try to better understand the confusing answer choice. Rephrasing is one way to tease out the sense of a confusing answer choice.

7. Your First Instinct

Many people struggle with multiple-choice tests because they overthink the questions. If you have studied sufficiently for the test, you should be prepared to trust your first instinct once you have carefully and completely read the question and all of the answer choices. There is a great deal of research suggesting that the mind can come to the correct conclusion very quickly once it has obtained all of the relevant information. At times, it may seem to you as if your intuition is working faster even than your reasoning mind. This may in fact be true. The knowledge you obtain while studying may be retrieved from your subconscious before you have a chance to work out the associations that support it. Verify your instinct by working out the reasons that it should be trusted.

8. Key Words

Many test takers struggle with multiple-choice questions because they have poor reading comprehension skills. Quickly reading and understanding a multiple-choice question requires a mixture of skill and experience. To help with this, try jotting down a few key words and phrases on a piece of scrap paper. Doing this concentrates the process of reading and forces the mind to weigh the relative importance of the question's parts. In selecting words and phrases to write down, the test taker thinks about the question more deeply and carefully. This is especially true for multiple-choice questions that are preceded by a long prompt.

9. Subtle Negatives

One of the oldest tricks in the multiple-choice test writer's book is to subtly reverse the meaning of a question with a word like *not* or *except*. If you are not paying attention to each word in the question, you can easily be led astray by this trick. For instance, a common question format is, "Which of the following is...?" Obviously, if the question instead is, "Which of the following is not...?," then the answer will be quite different. Even worse, the test makers are aware of the potential for this mistake and will include one answer choice that would be correct if the question were not negated or reversed. A test taker who misses the reversal will find what he or she believes to be a correct answer and will be so confident that he or she will fail to reread the question and discover the original error. The only way to avoid this is to practice a wide variety of multiple-choice questions and to pay close attention to each and every word.

10. Reading Every Answer Choice

It may seem obvious, but you should always read every one of the answer choices! Too many test takers fall into the habit of scanning the question and assuming that they understand the question because they recognize a few key words. From there, they pick the first answer choice that answers the question they believe they have read. Test takers who read all of the answer choices might discover that one of the latter answer choices is actually *more* correct. Moreover, reading all of the answer choices can remind you of facts related to the question that can help you arrive at the correct answer. Sometimes, a misstatement or incorrect detail in one of the latter answer choices will trigger your memory of the subject and will enable you to find the right answer. Failing to read all of the answer choices is like not reading all of the items on a restaurant menu: you might miss out on the perfect choice.

11. Spot the Hedges

One of the keys to success on multiple-choice tests is paying close attention to every word. This is never truer than with words like almost, most, some, and sometimes. These words are called "hedges" because they indicate that a statement is not totally true or not true in every place and time. An absolute statement will contain no hedges, but in many subjects, the answers are not always straightforward or absolute. There are always exceptions to the rules in these subjects. For this reason, you should favor those multiple-choice questions that contain hedging language. The presence of qualifying words indicates that the author is taking special care with his or her words, which is certainly important when composing the right answer. After all, there are many ways to be wrong, but there is only one way to be right! For this reason, it is wise to avoid answers that are absolute when taking a multiple-choice test. An absolute answer is one that says things are either all one way or all another. They often include words like *every*, *always*, *best*, and *never*. If you are taking a multiple-choice test in a subject that doesn't lend itself to absolute answers, be on your guard if you see any of these words.

12. Long Answers

In many subject areas, the answers are not simple. As already mentioned, the right answer often requires hedges. Another common feature of the answers to a complex or subjective question are qualifying clauses, which are groups of words that subtly modify the meaning of the sentence. If the question or answer choice describes a rule to which there are exceptions or the subject matter is complicated, ambiguous, or confusing, the correct answer will require many words in order to be expressed clearly and accurately. In essence, you should not be deterred by answer choices that seem excessively long. Oftentimes, the author of the text will not be able to write the correct answer without offering some qualifications and modifications. Your job is to read the answer choices thoroughly and completely and to select the one that most accurately and precisely answers the question.

13. Restating to Understand

Sometimes, a question on a multiple-choice test is difficult not because of what it asks but because of how it is written. If this is the case, restate the question or answer choice in different words. This process serves a couple of important purposes. First, it forces you to concentrate on the core of the question. In order to rephrase the question accurately, you have to understand it well. Rephrasing the question will concentrate your mind on the key words and ideas. Second, it will present the information to your mind in a fresh way. This process may trigger your memory and render some useful scrap of information picked up while studying.

14. True Statements

Sometimes an answer choice will be true in itself, but it does not answer the question. This is one of the main reasons why it is essential to read the question carefully and completely before proceeding to the answer choices. Too often, test takers skip ahead to the answer choices and look for true statements. Having found one of these, they are content to select it without reference to the question above. Obviously, this provides an easy way for test makers to play tricks. The savvy test taker will always read the entire question before turning to the answer choices. Then, having settled on a correct answer choice, he or she will refer to the original question and ensure that the selected answer is relevant. The mistake of choosing a correct-but-irrelevant answer choice is especially common on questions related to specific pieces of objective knowledge. A prepared test taker will have a wealth of factual knowledge at his or her disposal, and should not be careless in its application.

15. No Patterns

One of the more dangerous ideas that circulates about multiple-choice tests is that the correct answers tend to fall into patterns. These erroneous ideas range from a belief that B and C are the most common right answers, to the idea that an unprepared test-taker should answer "A-B-A-C-A-D-A-B-A." It cannot be emphasized enough that pattern-seeking of this type is exactly the WRONG way to approach a multiple-choice test. To begin with, it is highly unlikely that the test maker will plot the correct answers according to some predetermined pattern. The questions are scrambled and delivered in a random order. Furthermore, even if the test maker was following a pattern in the assignation of correct answers, there is no reason why the test taker would know which pattern he or she was using. Any attempt to discern a pattern in the answer choices is a waste of time and a distraction from the real work of taking the test. A test taker would be much better served by extra preparation before the test than by reliance on a pattern in the answers.

FREE DVD OFFER

Don't forget that doing well on your exam includes both understanding the test content and understanding how to use what you know to do well on the test. We offer a completely FREE Test Taking Tips DVD that covers world class test taking tips that you can use to be even more successful when you are taking your test.

All that we ask is that you email us your feedback about your study guide. To get your **FREE Test Taking Tips DVD**, email freedvd@studyguideteam.com with "FREE DVD" in the subject line and the following information in the body of the email:

- The title of your study guide.
- Your product rating on a scale of 1-5, with 5 being the highest rating.
- Your feedback about the study guide. What did you think of it?
- Your full name and shipping address to send your free DVD.

Introduction to the ISEE Middle Level Exam

Function of the Test

The Independent School Entrance Exam (ISEE) middle level is an entrance exam for grades 7 and 8 for independent schools across the U.S and abroad, developed and facilitated by the Educational Records Bureau (ERB). Complete scores are sent to the school of choice for application review. Although schools take the score into consideration, the score is not the only indication for acceptance.

Test Administration

The ISEE is offered in three separate testing seasons throughout the year through Prometric testing centers, approved ERB member school test sites, or ERB's New York City office. Testing seasons are Fall (August through November), Winter (December through March), and Spring/Summer (April through July). Students must register well before the test date, as registration closes two weeks before the test is given. For retesting, students may retake the ISEE up to 3 times in a 12-month admission period, but only once per testing season.

Test Format

Students must arrive at the testing center well before the test begins. Those who arrive late are not permitted into the testing room and must reschedule their exam. Personal items are not allowed in the exam room, including cell phones, paper, rulers, watches, compasses, etc. Paper testing permits students to bring four #2 pencils and two pens with either blue or black ink. Two ten-minute breaks are given during the exam. The first break is after the Quantitative Reasoning section. The second break is after the Mathematics Achievement section.

The ISEE middle level is for students entering seventh or eighth grade, and includes Verbal Reasoning, Quantitative Reasoning, Reading Comprehension, Mathematics Achievement, and an essay, in that order. Below is a table with more details on each section:

Section	Questions	Time
Verbal Reasoning	40	20 minutes
Quantitative Reasoning	37	35 minutes
Reading Comprehension	36	35 minutes
Mathematics Achievement	47	40 minutes
Essay	1 prompt	
Total		**160 minutes**

Scoring

Students should try to narrow down correct answers, as scores are determined by the number of questions answered correctly. There is no score difference between an omitted answer and an incorrect answer, and therefore no guessing penalty. Students are compared to fellow students by grade within the previous three years by a percentile ranking on the score report. For paper tests, scores are sent to the Individual Student Report (ISR) and an email will be sent when scores are ready. For online tests, score reports will post 3 to 5 days after the test. Essays are released only to ERB schools and are not scored.

Scoring comes with a raw score and a scaled score. Raw scores might be something like 35/40 for the Verbal Reasoning section, showing the amount of questions guessed correctly. Scaled scores have been converted to a higher number and show a range from 760 to 940.

Study Prep Plan for the ISEE Middle Level Exam

1 **Schedule -** Use one of our study schedules below or come up with one of your own.

2 **Relax -** Test anxiety can hurt even the best students. There are many ways to reduce stress. Find the one that works best for you.

3 **Execute -** Once you have a good plan in place, be sure to stick to it.

One Week Study Schedule

Day	Topic
Day 1	Verbal Reasoning
Day 2	Quantitative Reasoning
Day 3	Reading Comprehension
Day 4	Essay
Day 5	Practice Questions
Day 6	Answer Explanations
Day 7	Take Your Exam!

Two Week Study Schedule

Day	Topic	Day	Topic
Day 1	Verbal Reasoning	Day 8	Practice Questions
Day 2	Practice Questions	Day 9	Essay
Day 3	Quantitative Reasoning	Day 10	Practice Essay
Day 4	Practice Questions	Day 11	Review Practice Questions
Day 5	Reading Comprehension	Day 12	Review Answer Explanations
Day 6	Practice Questions	Day 13	(Study Break)
Day 7	Mathematics Achievement	Day 14	Take Your Exam!

One Month Study Schedule

Day	Topic	Day	Topic	Day	Topic
Day 1	Synonyms	Day 11	Quantitative Reasoning Practice Questions	Day 21	Practice Questions
Day 2	Sentence Completion	Day 12	Quantitative Reasoning Answer Explanations	Day 22	Answer Explanations
Day 3	Practice Questions	Day 13	Mathematics Achievement Practice Questions	Day 23	Parts of the Essay
Day 4	Answer Explanations	Day 14	Mathematics Achievement Answer Explanations	Day 24	The Short Overview
Day 5	Numbers and Operations	Day 15	Main Idea	Day 25	Applying Basic Knowledge of the Writing Process
Day 6	Algebra	Day 16	Supporting Ideas	Day 26	Developing a Well-Organized Paragraph
Day 7	Geometry	Day 17	Inferences	Day 27	Distinguishing Between Formal and Informal
Day 8	Measurement	Day 18	Vocabulary	Day 28	Tips for the Essay
Day 9	Data Analysis and Probability	Day 19	Organization/Logic	Day 29	(Study Break)
Day 10	Problem Solving	Day 20	Tone/Style/Figurative Language	Day 30	Take Your Exam!

Verbal Reasoning

Synonyms

This portion of the exam is made to test vocabulary skills. While logic and reasoning come into play in this section, they are not as important as the sentence completion questions. A prior knowledge of word meaning is helpful in order to answer correctly. If the meaning of the words is unknown, that's okay. Strategies should be used to rule out false answers and choose the correct ones. Here are some study strategies below.

Format of the Questions

The vocabulary questions are very simple. The prompt is just a single word. There are no special directions, alternate meanings, or analogies to work with. The objective is to find the meaning of the given word and then choose the answer that means the same thing or is closest in meaning to the given word. Note the example below:

Blustery

a. Hard
b. Windy
c. Mythical
d. Stoney

All of the vocabulary questions on the Verbal Reasoning section will appear exactly like the above sample. This is generally the standard layout throughout other exams, so some test takers may already be familiar with the structure. The principle remains the same. At the top of the section, clear directions will be given to choose the answer that best defines the given word. In this case, the answer is *windy* (B), since *windy* and *blustery* are synonymous.

Preparation

There is no set way to prepare for this portion of the exam that will guarantee a perfect score. This is because the words used on the test are unpredictable. There is no set list provided to study from. The definition of the provided word needs to be determined on the spot. This sounds challenging, but there are still ways to prepare mentally for the test. It may help to expand your vocabulary a little each day. Several resources are available, in books and online, that collect words and definitions that tend to show up frequently on standardized tests. Knowledge of words can increase the strength of your vocabulary.

Mindset is key. The meanings of challenging words can often be found by relying on the past experiences of the test-taker. How? Well, test-takers have been talking their entire lives—knowing words and how words work. It helps to have a positive mindset from the start. It's unlikely that all definitions of words will be known immediately, but the answer can still be found. There are aspects of words that are recognizable to help find the correct answers and eliminate the incorrect ones. Here are some of the factors that contribute to word meanings!

Word Origins and Roots

Studying a foreign language in school, particularly Latin or any of the romance languages (Latin-influenced), is helpful. English is a language highly influenced by Latin and Greek words. The roots of much of the English vocabulary have Latin origins; these roots can bind many words together and often point to a shared definition. Here's an example:

Fervent

a. Lame
b. Joyful
c. Thorough
d. Boiling

Fervent descends from the Latin word, *fervere*, which means "to boil or glow" and figuratively means "impassioned." The Latin root present in the word is *ferv*, which is what gives *fervent* the definition: showing great warmth and spirit or spirited, hot, glowing. This provides a link to *boiling* (D) just by root word association, but there's more to analyze. Among the other choices, none relate to *fervent*. The word *lame* (A) means crippled, disabled, weak, or inadequate. None of these match with *fervent*. While being *fervent* can reflect joy, *joyful* (B) directly describes "a great state of happiness," while *fervent* is simply expressing the idea of having very strong feelings—not necessarily joy. *Thorough* (C) means complete, perfect, painstaking, or with mastery; while something can be done thoroughly and fervently, none of these words match fervent as closely as *boiling* does. Not only does *boiling* connect in a linguistic way, but also in the way it is used in our language. While *boiling* can express being physically hot and undergoing a change, *boiling* is also used to reflect emotional states. People say they are "boiling over" when in heighted emotional states; "boiling mad" is another expression. *Boiling*, like *fervent*, also embodies a sense of heightened intensity. This makes *boiling* the best choice!

The Latin root *ferv* is seen in other words such as *fervor, fervid*, and even *ferment*. All of them are connected to and can be described by boil or glow, whether it is in a physical sense or in a metaphorical one. Such a pattern can be seen in other word sets! Here's another example:

Gracious

a. Fruitful
b. Angry
c. Grateful
d. Understood

This one's a little easier; the answer is grateful (*C*) because both words mean thankful! Even if the meanings of both words are known, there's a connection found by looking at the beginnings of both words: *gra/grat*. Once again, there is a root that stretches back to classical language. Both terms come from the Latin, *gratis*, which literally means "thanks."

Understanding root words can help identify the meaning in a lot of word choices and help the test-taker grasp the nature of the given word. Many dictionaries, both in book form and online, offer information on the origins of words, which highlight these roots. When studying for the test, it helps to look up an unfamiliar word for its definition and then check to see if it has a root that can be connected to any of the other terms.

Pay Attention to Prefixes

The prefix of a word reveals a lot about its definition. A **prefix** is a set of letters at the beginning of a word. When encountering any unfamiliar words, try looking at prefixes to discern the definition and then compare that with the choices. The prefix should be determined to help find the word's meaning. Here's an example question:

Premeditate

a. Sporadic
b. Calculated
c. Interfere
d. Determined

With *premeditate*, there's the common prefix *pre*. This helps draw connections to other words like *prepare* or *preassemble*. *Pre* refers to "before, already being, or having already." *Meditate* means to think or plan. *Premeditate* means to think or plan beforehand with intent. Therefore, a term that deals with thinking or planning should be found, but also something done in preparation. Among the word choices, *determined* (D) is an adjectives with no hint of being related to something done before or in preparation. This choice is incorrect. *Sporadic* (A) refers to events happening in irregular patterns, so this is quite the opposite of *premeditated*. *Interfere* (C) also has nothing to do with *premeditate*; it goes counter to *premeditate* in a way similar to *sporadic*. *Calculated* (B), however, fits! A route and the cost of starting a plan can be calculated. *Calculated* refers to acting with a full awareness of consequences, so inherently planning is involved. In fact, *calculated* is synonymous with *premeditated*, thus making it the correct choice. Just by paying attention to a prefix, the doors to a meaning can open to help easily figure out which word would be the best choice. Here's another example:

Regain

a. Erupt
b. Ponder
c. Seek
d. Recoup

Recoup (D) is the correct answer. The prefix *re* often appears in front of words to give them the meaning of occurring again. *Regain* means to repossess something that was lost. *Recoup*, which also has the *re* prefix, literally means to regain. In this example, both the given word and the answer share the *re* prefix, which makes the pair easy to connect. However, don't rely only on prefixes to choose an answer. Make sure to analyze all options before marking an answer. Going through the other words in this sample, none of them come close to meaning regain except *recoup*. After checking to make sure that *recoup* is the best matching word, then mark it!

Positive Versus Negative Sounding Words

Another tool for the mental toolbox is simply distinguishing whether a word has a positive or negative connotation. Like electrical wires, words carry energy; they are crafted to draw certain attention and to have certain strength to them. Words can be described as **positive** and uplifting (a stronger word) or they can be **negative** and scathing (a stronger word). Sometimes they are **neutral**—having no particular connotation. Distinguishing how a word is supposed to be interpreted will not only help learn its definition, but also draw parallels with word choices. While it's true that words must usually be taken in the context of how they are used, word definitions have inherent meanings as well. So, they have a distinct vibe to pick up on. Here is an example:

Excellent

a. Fair
b. Optimum
c. Reasonable
d. Agitation

As you know, *excellent* is a very positive word. It refers to something being better than good, above average. In this sample, *agitation* (D) can be eliminated because it has a negative connotations. *Reasonable* (C) is more or less a neutral word: it's not bad but it doesn't communicate the higher quality that *excellent* represents. It's just, well, *reasonable*. This leaves the possible choices of *fair* (A) and *optimum* (B). Or does it? Fair *is* a positive word; it's used to describe things that are good, even beautiful. But in the modern context, *fair* is defined as good, but somewhat average or just decent: "You did a fairly good job." or, "That was fair." On the other hand, *optimum* is positive and is a stronger word. *Optimum* describes the most favorable outcome. This makes *optimum* the best word choice that matches *excellent* in both strength and connotation. Not only are the two words positive, but they also express the same level of positivity! Here's another example:

Repulse

a. Draw
b. Encumber
c. Magnify
d. Disgust

Repulse sounds negative when said. It is commonly used in the context of something being repulsive, disgusting, or that which is distasteful. It's also defined as an attack that drives people away. This tells us we need a word that also carries a negative meaning. *Magnify* (C) is positive, while *draw* (A) is neutral. *Encumber* (B) and *disgust (*D) are negative. *Disgust* is a stronger negative than *encumber.* Of all the words given, only *disgust* directly defines a feeling of distaste and aversion that is synonymous with *repulse* and matches in both negativity and strength.

Parts of Speech

It is often very helpful to determine the **part of speech** of a word. Is it an adjective, adverb, noun, or verb? Often the correct answer will also be the same part of speech as the given word. Isolate the part of speech and what it describes and look for an answer choice that also describes the same part of speech. For example: if the given word is an adverb describing an action word, then look for another adverb describing an action word.

Swiftly

a. Fast
b. Quietly
c. Quickly
d. Sudden

Swiftly is an adverb that describes the speed of an action. *Fast* (A) and **sudden** (D) can be eliminated because they are not adverbs, and *quietly* (B) can be eliminated because it does not describe speed. This leaves *quickly* (C), which is the correct answer. *Fast* and *sudden* may throw off some test takers because they both describe speed, but *quickly* matches more closely because it is an adverb, and *swiftly* is also an adverb.

Place the Word in a Sentence

Often it is easier to discern the meaning of a word if it is used in a sentence. If the given word can be used in a sentence, then try replacing it with some of the answer choices to see which words seem to make sense in the same sentence. Here's an example:

Remarkable

a. Often
b. Capable
c. Outstanding
d. Shining

A sentence can be formed with the word *remarkable*. "My grade point average is remarkable." None of the examples make sense when replacing the word *remarkable* in the sentence other than the word *outstanding* (C), so *outstanding* is the obvious answer. *Shining* (D) is also a word with a positive connotation, but *outstanding* fits better in the sentence.

Pick the Closest Answer

As the answer choices are reviewed, two scenarios might stand out. An exact definition match might not be found for the given word among the choices, or there are several word choices that can be considered synonymous to the given word. This is intentionally done to test the ability to draw parallels between the words to produce an answer that best fits the prompt word. Again, the closest fitting word will be the answer. Even when facing these two circumstances, finding the one word that fits best is the proper strategy. Here's an example:

Insubordination

a. Cooperative
b. Disciplined
c. Rebel
d. Contagious

Insubordination refers to a defiance or utter refusal of authority. Looking over the choices, none of these terms provide definite matches to insubordination like *insolence, mutiny*, or *misconduct* would. This is fine; the answer doesn't have to be a perfect synonym. The choices don't reflect insubordination in any way, except *rebel* (C). After all, when *rebel* is used as a verb, it means to act against authority. It's also used as a noun: someone who goes against authority. Therefore, *rebel* is the best choice.

As with the verbal analogies section, being a detective is a good course to take. Two or even three choices might be encountered that could be the answer. However, the answer that most perfectly fits the prompt word's meaning is the best answer. Choices should be narrowed one word at a time. The least-connected word should be eliminated first and then proceed until one word is left that is the closest synonym.

Sequence

a. List
b. Range
c. Series
d. Replicate

A *sequence* reflects a particular order in which events or objects follow. The two closest options are *list* (A) and *series* (C). Both involve grouping things together, but which fits better? Consider each word more carefully. A list is comprised of items that fit in the same category, but that's really it. A list doesn't have to follow any particular order; it's just a list. On the other hand, a series is defined by events happening in a set order. A series relies on sequence, and a sequence can be described as a series. Thus, *series* is the correct answer!

Sentence Completion

Parts of Speech

Also referred to as word classes, **parts of speech** refer to the various categories in which words are placed. Words can be placed in any one or a combination of the following categories:

- Nouns
- Determiners
- Pronouns
- Verbs
- Adjectives
- Adverbs
- Prepositions
- Conjunctions

Understanding the various parts of speech used in the English language helps readers to better understand the written language.

Nouns

Nouns are defined as any word that represents a person, place, animal, object, or idea. Nouns can identify a person's title or name, a person's gender, and a person's nationality, such as *banker, commander-in-chief, female, male*, or *American*.

With animals, nouns identify the kingdom, phylum, class, etc. For example: animal is *elephant*; phylum is *chordata*; or class is *mammalia*.

When identifying places, nouns refer to a physical location, a general vicinity, or the proper name of a city, state, or country. Some examples include the *desert,* the *East, Phoenix, Arizona*, or the *United States*.

There are eight types of nouns: common, proper, countable, uncountable, concrete, abstract, compound, and collective.

Common nouns are used in general terms, without specific identification. Examples include *girl, boy, country*, or *school*. **Proper nouns** refer to the proper name given to people, places, animals, or entities, such as *Melissa, Martin, Italy*, or *Harvard*.

Countable nouns can be counted: *one* car, *two* cars, or *three* cars. **Uncountable nouns** cannot be counted, such as *air, liquid*, or *gas*.

To be abstract is to exist, but only in thought or as an idea. An **abstract noun** cannot be physically touched, seen, smelled, heard, or tasted. These include *chivalry, day, fear, thought, truth, friendship*, or *freedom*.

To be **concrete** is to be seen, touched, tasted, heard, and/or smelled. Examples include *pie, snow, tree, bird, desk, hair*, or *dog*.

A **compound noun** is another term for an open compound word. Any noun that is written as two nouns that together form a specific meaning is a compound noun, such as *post office, ice cream*, or *swimming pool*.

A **collective noun** is formed by grouping three or more words to create one meaning. Examples include *bunch of flowers, herd of elephants, flock of birds*, or *school of fish*.

Determiners

Determiners modify a noun and usually refer to something specific. Determiners fall into one of four categories: articles, demonstratives, quantifiers, or possessive determiners.

Articles can be both **definite** articles, as in *the*, and **indefinite** as in *a, an, some,* or *any*:

> *The* man with *the* red hat.
>
> A flower growing in *the* yard.
>
> *Any* person who visits *the* store.

There are four different types of **demonstratives**: *this, that, these*, and *those*.

True demonstrative words will not directly precede the noun of the sentence but will *be* the noun. Some examples:

> *This* is the one.
>
> *That* is the place.
>
> *Those* are the files.

Once a demonstrative is placed directly in front of the noun, it becomes a demonstrative pronoun:

> *This* one is perfect.
>
> *That* place is lovely.
>
> *Those* boys are annoying.

Quantifiers proceed nouns to give additional information for how much or how many. They can be used with countable and uncountable nouns:

She bought *plenty* of apples.

Few visitors came.

I got a *little* change.

Possessive determiners, sometimes called possessive adjectives, indicate possession. They are the possessive forms of personal pronouns, such as *for my, your, hers, his, its, their*, or *our*:

That is *my* car.

Tom rode *his* bike today.

Those papers are *hers.*

Pronouns

Pronouns are words that stand in place of nouns. There are three different types of pronouns: subjective pronouns (*I, you, he, she, it, we, they*), objective pronouns (*me, you, him, her it, us, them*), and possessive pronouns (*mine, yours, his, hers, ours, theirs*).

You'll see some words are found in more than one pronoun category. See examples and clarifications below:

You are going to the movies.

You is a subjective pronoun; it is the subject of the sentence and is performing the action.

I threw the ball to *you.*

You is an objective pronoun; it is receiving the action and is the object of the sentence.

We saw *her* at the movies.

Her is an objective pronoun; it is receiving the action and is the object of the sentence.

The house across the street from the park is *hers.*

Hers is a possessive pronoun; it shows possession of the house and is used as a possessive pronoun.

Verbs

Verbs are words in a sentence that show action or state. Just as there can be no sentence without a subject, there can be no sentence without a verb. In these sentences, notice verbs in the present, past, and future tenses. To form some tenses in the future and in the past requires **auxiliary**, or **helping**, verbs:

I *see* the neighbors across the street.

See is an action.

We *were eating* at the picnic.

Eating is the main action, and the verb *were* is the past tense of the verb *to be*, and is the helping or auxiliary verb that places the sentence in the past tense.

> You *will turn* 20 years of age next month.

Turn is the main verb, but to show a future tense, *will* is the helping verb to show future tense of the verb *to be*.

Adjectives

Adjectives are a special group of words used to modify or describe a noun. Adjectives provide more information about the noun they modify. For example:

> The boy went to school. (There is no adjective.)

Rewrite the sentence, adding an adjective to further describe the boy and/or the school:

> The *young* boy went to the *old* school. (The adjective *young* describes the boy, and the adjective *old* describes the school.)

Adverbs

Adverb can play one of two roles: to modify the adjective or to modify the verb. For example:

> The young boy went to the old school.

We can further describe the adjectives *young* and *old* with adverbs placed directly in front of the adjectives:

> The *very* young boy went to the *very* old school. (The adverb *very* further describes the adjectives *young* and *old*.)

Other examples of using adverbs to further describe verbs:

> The boy *slowly* went to school.
>
> The boy *quickly* went to school.

The adverbs *slowly* and *quickly* further modify the verbs.

Prepositions

Prepositions are special words that generally precede a noun. Prepositions clarify the relationship between the subject and another word or element in the sentence. They clarify time, place, and the positioning of subjects and objects in a sentence. Common prepositions in the English language include: *near, far, under, over, on, in, between, beside, of, at, until, behind, across, after, before, for, from, to, by,* and *with*.

Conjunctions

Conjunctions are a group of unique words that connect clauses or sentences. They also work to coordinate words in the same clause. It is important to choose an appropriate conjunction based on the meaning of the sentence. Consider these sentences:

I really like the flowers, *however* the smell is atrocious.

I really like the flowers, *besides* the smell is atrocious.

The conjunctions *however* and *besides* act as conjunctions, connecting the two ideas: *I really like the flowers,* and, *the smell is atrocious*. In the second sentence, the conjunction *besides* makes no sense and would confuse the reader. You must consider the message you wish to convey and choose conjunctions that clearly state that message without ambiguity.

Some conjunctions introduce an opposing opinion, thought, or fact. They can also reinforce an opinion, introduce an explanation, reinforce cause and effect, or indicate time. For example:

She wishes to go to the movies, *but* she doesn't have the money. (Opposition)

The professor became ill, *so* the class was postponed. (Cause and effect)

They visited Europe *before* winter came. (Time)

Each conjunction serves a specific purpose in uniting two separate ideas. Below are common conjunctions in the English language:

Opposition	Cause & Effect	Reinforcement	Time	Explanation
however	therefore	besides	afterward	for example
nevertheless	as a result	anyway	before	in other words
but	because of this	after all	firstly	for instance
although	consequently	furthermore	next	such as

Prefixes and Suffixes

In this section, we will look at the meaning of various prefixes and suffixes when added to a root word. A **prefix** is a combination of letters found at the beginning of a word. A **suffix** is a combination of letters found at the end. A **root word** is the word that comes after the prefix, before the suffix, or between them both. Sometimes a root word can stand on its own without either a prefix or a suffix. More simply put:

Prefix + Root Word = Word

Root Word + Suffix = Word

Prefix + Root Word + Suffix = Word

Root Word = Word

Knowing the definitions of common prefixes and suffixes is helpful when you are trying to find out the meaning of a word you don't know. Also, knowing prefixes can help you find out the number of things, the negative of something, or the time and space of an object. Understanding suffixes can help when trying to find out the meaning of an adjective, noun, or verb.

The following charts look at some of the most common prefixes, what they mean, and how they're used to find out a word's meaning:

Number and Quantity Prefixes

Prefix	**Definition**	**Example**
bi-	two	bicycle, bilateral
mono-	one, single	monopoly, monotone
poly-	many	polygamy, polygon
semi-	half, partly	semiannual, semicircle
uni-	one	unicycle, universal

Here's an example of a number prefix:

The girl rode on a *bicycle* to school.

Look at the word *bicycle*. The root word (*cycle*) comes from the Greek and means *wheel*. The prefix *bi-* means *two*. The word *bicycle* means two wheels! When you look at any bicycle, they all have two wheels. If you had a unicycle, your bike would only have one wheel, because *uni-* means *one*.

Negative Prefixes

Prefix	**Definition**	**Example**
a-	without, lack of	amoral, atypical
in-	not, opposing	inability, inverted
non-	not	nonexistent, nonstop
un-	not, reverse	unable, unspoken

Here's an example of a negative prefix:

The girl was *insensitive* to the boy who broke his leg.

Look at the word *insensitive.* In the chart above, the prefix *in-* means *not* or *opposing*. Replace the prefix with *not*. Now place *not* in front of the word *sensitive.* Now we see that the girl was "not sensitive" to the boy who broke his leg. In simpler terms, she showed that she did not care. These are easy ways to use prefixes and suffixes in order to find out what a word means.

Time and Space Prefixes

Prefix	Definition	Example
a-	in, on, of, up, to	aloof, associate
ab-	from, away, off	abstract, absent
ad-	to, towards	adept, adjacent
ante-	before, previous	antebellum, antenna
anti-	against, opposing	anticipate, antisocial
cata-	down, away, thoroughly	catacomb, catalogue
circum-	around	circumstance, circumvent
com-	with, together, very	combine, compel
contra-	against, opposing	contraband, contrast
de-	from	decrease, descend
dia-	through, across, apart	diagram, dialect
dis-	away, off, down, not	disregard, disrespect
epi-	upon	epidemic, epiphany
ex-	out	example, exit
hypo-	under, beneath	hypoallergenic, hypothermia
inter-	among, between	intermediate, international
intra-	within	intrapersonal, intravenous
ob-	against, opposing	obtain, obscure
per-	through	permanent, persist
peri-	around	periodontal, periphery
post-	after, following	postdate, postoperative
pre-	before, previous	precede, premeditate
pro-	forward, in place of	program, propel
retro-	back, backward	retroactive, retrofit
sub-	under, beneath	submarine, substantial
super-	above, extra	superior, supersede
trans-	across, beyond, over	transform, transmit
ultra-	beyond, excessively	ultraclean, ultralight

Here's an example of a space prefix:

> The teacher's motivational speech helped *propel* her students toward greater academic achievement.

Look at the word *propel.* The prefix *pro-* means *forward. Forward* means something related to time and space. *Propel* means to drive or move in a forward direction. Therefore, knowing the prefix *pro-* helps interpret that the students are moving forward *toward greater academic achievement*.

Miscellaneous Prefixes

Prefix	Definition	Example
belli-	war, warlike	bellied, belligerent
bene-	well, good	benediction, beneficial
equi-	equal	equidistant, equinox
for-	away, off, from	forbidden, forsaken
fore-	previous	forecast, forebode
homo-	same, equal	homogeneous, homonym
hyper-	excessive, over	hyperextend, hyperactive
in-	in, into	insignificant, invasive
magn-	large	magnetic, magnificent
mal-	bad, poorly, not	maladapted, malnourished
mis-	bad, poorly, not	misplace, misguide
mor-	death	mortal, morgue
neo-	new	neoclassical, neonatal
omni-	all, everywhere	omnipotent, omnipresent
ortho-	right, straight	orthodontist, orthopedic
over-	above	overload, overstock,
pan-	all, entire	panacea, pander
para-	beside, beyond	paradigm, parameter
phil-	love, like	philanthropy, philosophic
prim-	first, early	primal, primer
re-	backward, again	reload, regress
sym-	with, together	symmetry, symbolize
vis-	to see	visual, visibility

Here's another prefix example:

> The computer was *primitive*; it still had a floppy disk drive!

The word *primitive* has the prefix *prim-*. The prefix *prim-*indicates being *first* or *early*. *Primitive* refers to the historical development of something. Therefore, the sentence is saying that the computer is an older model, because it still has a floppy disk drive.

The charts that follow review some of the most common suffixes. They also include examples of how the suffixes are used to determine the meaning of a word. Remember, suffixes are added to the *end* of a root word:

Adjective Suffixes

Suffix	Definition	Example
-able (-ible)	capable of being	teachable, accessible
-esque	in the style of, like	humoresque, statuesque
-ful	filled with, marked by	helpful, deceitful
-ic	having, containing	manic, elastic
-ish	suggesting, like	malnourish, tarnish
-less	lacking, without	worthless, fearless
-ous	marked by, given to	generous, previous

Here's an example of an adjective suffix:

> The live model looked so *statuesque* in the window display; she didn't even move!

Look at the word *statuesque*. The suffix *-esque* means *in the style of* or *like*. If something is *statuesque,* it's *like a statue*. In this sentence, the model looks like a statue.

Noun Suffixes

Suffix	**Definition**	**Example**
-acy	state, condition	literacy, legacy
-ance	act, condition, fact	distance, importance
-ard	one that does	leotard, billiard
-ation	action, state, result	legislation, condemnation
-dom	state, rank, condition	freedom, kingdom
-er (-or)	office, action	commuter, spectator
-ess	feminine	caress, princess
-hood	state, condition	childhood, livelihood
-ion	action, result, state	communion, position
-ism	act, manner, doctrine	capitalism, patriotism
-ist	worker, follower	stylist, activist
-ity (-ty)	state, quality, condition	community, dirty
-ment	result, action	empowerment, segment
-ness	quality, state	fitness, rudeness
-ship	position	censorship, leadership
-sion (-tion)	state, result	tension, transition
-th	act, state, quality	twentieth, wealth
-tude	quality, state, result	attitude, latitude

Look at the following example of a noun suffix:

> The *spectator* cheered when his favorite soccer team scored a goal.

Look at the word *spectator*. The suffix *-or* means *action*. In this sentence, the *action* is to *spectate* (watch something). Therefore, a *spectator* is someone involved in watching something.

Verb Suffixes

Suffix	**Definition**	**Example**
-ate	having, showing	facilitate, integrate
-en	cause to be, become	frozen, written
-fy	make, cause to have	modify, rectify
-ize	cause to be, treat with	realize, sanitize

Here's an example of a verb suffix:

> The preschool had to *sanitize* the toys every Tuesday and Thursday.

In the word *sanitize*, the suffix *-ize* means *cause to be* or *treat with*. By adding the suffix *-ize* to the root word *sanitary*, the meaning of the word becomes active: *cause to be sanitary*.

Practice Questions

Synonyms

Select the word that is closest in meaning to the original word.

1. TEXTILE
    a. Fabric
    b. Document
    c. Knit
    d. Ornament

2. OFFSPRING
    a. Bounce
    b. Parent
    c. Music
    d. Child

3. PERMIT
    a. Law
    b. Parking
    c. Allow
    d. Jail

4. PERPLEXED
    a. Annoyed
    b. Vengeful
    c. Injured
    d. Confused

5. ROTATION
    a. Wheel
    b. Year
    c. Spin
    d. Orbit

6. CONSISTENT
    a. Steady
    b. Contains
    c. Sticky
    d. Texture

7. PRINCIPLE
    a. Principal
    b. Leader
    c. President
    d. Foundation

8. PERIMETER
    a. Outline
    b. Area
    c. Side
    d. Volume

9. SYMBOL
    a. Text
    b. Music
    c. Clang
    d. Emblem

10. GERMINATE
    a. Grow
    b. Sick
    c. Infect
    d. Plants

11. DEDUCE
    a. Explain
    b. Win
    c. Reason
    d. Gamble

12. ELUCIDATE
    a. Learn
    b. Enlighten
    c. Plan
    d. Corroborate

13. VERIFY
    a. Criticize
    b. Change
    c. Teach
    d. Substantiate

14. IRATE
    a. Anger
    b. Knowledge
    c. Taciturn
    d. Confused

15. REGALE
    a. Remember
    b. Bore
    c. Outnumber
    d. Entertain

16. WEARY
    a. Tired
    b. Clothing
    c. Happy
    d. Whiny

17. VAST
    a. Rapid
    b. Expansive
    c. Small
    d. Ocean

18. DEMONSTRATE
    a. Tell
    b. Show
    c. Build
    d. Make

19. ORCHARD
    a. Flower
    b. Fruit
    c. Grove
    d. Farm

20. ALLUDE
    a. Organize
    b. Interpret
    c. Refer
    d. Grasp

Sentence Completion

Select the word or phrase that most correctly completes the sentence.

21. The students were saying nice things about their friend Michael that __________ his reputation.
    a. boosted
    b. removed
    c. erased
    d. disgraced

22. The frightened coyote tried to escape the terrifying cougar by backing up near the edge of a(n) __________.
    a. precipice
    b. dagger
    c. atmosphere
    d. compliment

23. Even after eating three full meals and several snacks, the athlete's hunger was not __________.
    a. broken
    b. perilous
    c. affordable
    d. satisfied

24. Alison Creek, a waterway located in California, could not be used as a source of drinking water because of its __________ quality.
    a. beneficial
    b. admirable
    c. detrimental
    d. unique

25. Josephine did not have much time to complete her test, so she started concentrating on the main topics instead of on the __________ material.
    a. important
    b. necessary
    c. irrelevant
    d. critical

26. Juan is a talented painter best known for his use of __________.
    a. hyperbole
    b. color
    c. onomatopoeia
    d. simile

27. Chrissy is a great team leader because she is able to __________ her team to produce rapid results.
    a. inspire
    b. disable
    c. confuse
    d. daunt

28. The students were so exhausted after the week of practice testing that they acted __________ when asked to perform one more test.
    a. animated
    b. energized
    c. motivated
    d. sluggish

29. The music tonight was so __________ that I had to cover my ears to dilute the sound.
    a. serene
    b. placid
    c. rowdy
    d. muffled

30. The thief __________ with the purse before anyone knew what had happened.
    a. bolted
    b. lagged
    c. slowed
    d. waned

31. The teacher recognized the immature writing style of the first grader because his papers used __________ language.
    a. excellent
    b. mediocre
    c. prestige
    d. perfect

32. Much to the __________ of the unprepared students, the teacher called on students randomly to deliver unscripted speeches.
    a. assurance
    b. dismay
    c. elation
    d. pleasure

33. Margaret moved into a new house that was __________ and extravagant.
    a. destitute
    b. scanty
    c. puny
    d. fancy

34. Rachel wanted __________ for the time she had spent going through orientation.
    a. dedication
    b. escalation
    c. compensation
    d. accumulation

35. The __________ of the river was astounding; the group could see all the way down to the bottom.
    a. clarity
    b. murkiness
    c. shadow
    d. gloom

36. At one time, the Roman Empire was one of the most __________ military, economic, political, and cultural forces in the world.
    a. lax
    b. robust
    c. disappointing
    d. demure

37. Since the tadpole had a __________ outer covering, the biologist could not tell what the internal organs looked like.
    a. sheer
    b. clear
    c. solid
    d. bright

38. Ella's personality was so __________; she got invited to all the birthday parties and the other students loved being around her.
    a. fickle
    b. headstrong
    c. dejected
    d. engaging

39. The practice was so __________, we were too tired to play the game the next day.
    a. strenuous
    b. passive
    c. malicious
    d. gracious

40. Carley was __________ during the movie because she had a similar dog years before who had passed away.
    a. indifferent
    b. sentimental
    c. indignant
    d. animated

Answer Explanations

Synonyms

1. A: A *textile* is another word for a *fabric*. The most confusing alternative choice in this case is *knit*, because some textiles are knit, but *textile* and *knit* are not synonyms and plenty of textiles are not knit.

2. D: *Offspring* are the children of parents. This word is especially common when talking about the animal kingdom, although it can be used with humans as well. *Offspring* does have the word *spring* in it, although it has nothing to do with bouncing or jumping. The other answer choice, *parent*, may be somewhat tricky because parents have offspring, but for this reason, they are not synonyms.

3. C: *Permit* can be a verb or a noun. As a verb, it means to allow or give authorization for something. As a noun, it generally refers to a document or something that has been authorized, like a parking permit or driving permit, allowing the authorized individual to park or drive under the rules of the document.

4. D: *Perplexed* means baffled or puzzled, which are synonymous with *confused.*

5. C: *Rotation* means to *spin* or turn, such as a wheel rotating on a car, although wheel does not mean rotation.

6. A: Something that is consistent is *steady*, predictable, reliable, or constant. The tricky one here is that the word *consistency* comes from the word consistent and may describe a text or something that is sticky. *Consistent* also comes from the word *consist,* which means to *contain* (Choice *B*). Test takers must be discerning readers and knowledgeable about vocabulary to recognize the difference in these words and look for the true synonym of *consistent.*

7. D: A *principle* is a *foundation* or a guiding idea or belief. Someone with good moral character is described as having strong principles. Test takers must be careful not to get confused with the homonyms *principle* and *principal*, because these words have very different meanings. A *principal* is the leader of a school, and the word *principal* also refers to the main or most important idea or thing.

8. A: *Perimeter* refers to the outline or borders of an object. Test takers may recognize that word from math class, where perimeter refers to the edges or distance around an enclosed shape. Some of the other answer choices refer to other math vocabulary encountered in geometry lessons, but do not have the same meaning as *perimeter.*

9. D: A *symbol* is an object, picture, or sign that is used to represent something. For example, a pink ribbon is a symbol for breast-cancer awareness and a flag can be a symbol for a country. The tricky part of this question was also knowing the meaning of *emblem,* which typically describes a design that represents a group or concept, much like a symbol. Emblems often appear on flags or a coat of arms.

10. A: *Germinate* means to develop or *grow* and most often refers to sprouting seeds as a new plant first breaks through the seed coat. It can also refer to the development of an idea. Choice *D* may be an attractive choice since plants germinate, but germinate does not mean *plant.*

11. C: To *deduce* something is to figure it out using reasoning. Although this might cause a *win* and prompt an *explanation* to further understanding, the art of deduction is logical reasoning.

12. B: To *elucidate*, a light is figuratively shined on a previously unknown or confusing subject. This Latin root, "lux" meaning "light," prominently figures into the solution. *Enlighten* means to educate or bring into the light.

13. D: Looking at the Latin word "veritas," meaning "truth," will yield a clue as to the meaning of *verify*. To *verify* is the act of finding or assessing the truthfulness of something. This usually means amassing evidence to *substantiate* a claim. *Substantiate* means to provide evidence to prove a point.

14. A: *Irate* means being in a state of *anger*. Clearly this is a negative word that can be paired with another word in kind. The closest word to match this is *anger*. Research would also reveal that *irate* comes from the Latin "*ira*," which means anger.

15. D: *Regale* means to amuse someone with a story. This is a very positive word; the best way to eliminate choices is to look for a term that matches regale in both positive context and definition. *Entertain* is both a positive word and a synonym of *regale*.

16. A: *Weary* most closely means *tired*. Someone who is *weary* and *tired* may be *whiny*, but they do not necessarily mean the same thing.

17. B: Something that is *vast* is far-reaching and *expansive*. Choice *D*, *ocean*, may be described as vast, but the word alone doesn't mean vast. The heavens or skies may also be described as vast. Someone's imagination or vocabulary can also be vast.

18. B: To *demonstrate* something means to *show* it. A *demonstration* is a modeling or show-and-tell type of example that is usually visual.

19. C: An *orchard* is most like a *grove* because both are areas like plantations that grow different kinds of fruit. Many citrus fruits are grown in groves, but either word can be used to describe many fruit-bearing trees in one area. Choice *D*, *farm*, may have an orchard or grove on the property, but they are not the same thing, and many farms do not grow fruit trees.

20. C: *Allude* means to refer to something indirectly, so *refer* is the best answer here. *Organize*, *Interpret*, and *grasp* are incorrect answers.

Sentence Completion

21. A: We can assume that the students were saying nice things about their friend and that it *boosted* the friend's reputation. The other words mean being dragged down, so they are incorrect.

22. A: The *precipice* is a brink of a rock or mountain, so this is the correct answer choice. The other answer choices do not fit the context of the sentence.

23. D: The athlete's hunger was not *satisfied* even after eating three full meals and several snacks. *Broken*, *perilous*, and *affordable* are not characteristics of hunger, so these are incorrect.

24. C: The word *detrimental* means damaging or unhealthy. Therefore, if the water cannot be used as a source to drink from, it must be unhealthy, or *detrimental*, to someone's health.

25. C: We can rule Choices *A*, *B*, and *D* out. *Important, necessary,* and *crucial* all mean that the material is valuable to Josephine. However, from the context of the sentence, we should choose something that means the material besides the main topic is unnecessary. Choice *C*, *irrelevant*, means unimportant or unnecessary, so this is the best choice.

26. B: Choices *A*, *C*, and *D*, *hyperbole, onomatopoeia,* and *simile*, are all terms used in writing. Choice *B*, *color*, is specific to painting.

27. A: Chrissy is supposed to be a great team leader, so let's look at the words that are positive rather than negative. If someone *inspires* someone else then they are leading that person to encouragement and support, so *inspire* is the best answer. *Disable, confuse*, and *daunt* all denote negative actions, so these are not characteristics of a great team leader.

28. D: The students would act *sluggish*, or tired, when asked to perform one more test because of their exhaustion. The other words mean having energy, so these are incorrect.

29. C: The music was *rowdy*, which means loud or boisterous enough for someone to have to cover their ears. The other words denote calm or quiet, so they are incorrect.

30. A: *Bolted* means to run away quickly, so this is the best answer for someone who is stealing a purse.

31. B: *Mediocre* means average or inferior, so this is the best answer choice. The other words mean superiority, so they are incorrect, since an immature writing style means that the paper would not be superior.

32. B: The word *dismay* means being disappointed or showing stress, so this is the best word to use. Students who are supposed to give an unprepared speech will probably feel nervous about the situation, so the other answer choices, which all mean confident or happy, are incorrect.

33. D: Let's look for words that are most closely related to the word *extravagant*. Choice *D*, *fancy*, means extravagant and ornamental, so these are the best answers. Choices *A*, *B*, and *C* are versions of the words poor or cheap, so these are incorrect.

34. C: *Compensation* is the best word to put here because it means to repay or to make up for. If Rachel is sitting through an orientation for her job, she probably expects to be paid for that time she is learning.

35. A: The word *clarity* fits best here. Let's look at the context of the sentence. The second part of the sentence tells us that the group could see all the way down to the bottom of the river. This means that the water in the river must have been *clear*. *Clarity* means clearness, so this is our best answer choice.

36. B: *Robust* is the best word here because it means strong and healthy. A "cultural force" would be one with strong military, economic, and political power.

37. C: The word *solid* is the best fit here. If something has a solid outer surface, it would be impossible to view the inner organs.

38. D: *Engaging* means charming or attractive. People who have engaging personalities are usually very well-liked by definition, so this is the best answer. *Fickle* means unreliable. *Headstrong* means stubborn. *Dejected* means sad or depressed.

39. A: The best word to complete the sentence is *strenuous*, which means requiring lots of hard work. A strenuous practice would have tired the players out. *Passive* means lacking in energy. *Malicious* means intending to cause harm. *Gracious* means charming or generous.

40. B: *Sentimental* means overly emotional, especially in regard to a past feeling state. The word *indifferent* means showing no interest or concern. *Indignant* means outraged. *Animated* means lively or spirited.

Quantitative Reasoning

Numbers and Operations

Basic Operations of Arithmetic

There are four different basic operations used with numbers: addition, subtraction, multiplication, and division.

Addition

Addition is the combination of two numbers, so their quantities are added together cumulatively. The sign for an addition operation is the + symbol. For example, $9 + 6 = 15$. The 9 and 6 combine to achieve a cumulative value, called a **sum**.

Addition holds the commutative property, which means that the order of the numbers in an addition equation can be switched without altering the result. The formula for the commutative property is $a + b = b + a$. Let's look at a few examples to see how the commutative property works:

$$7 = 3 + 4 = 4 + 3 = 7$$

$$20 = 12 + 8 = 8 + 12 = 20$$

Addition also holds the **associative property**, which means that the grouping of numbers doesn't matter in an addition problem. In other words, the presence or absence of parentheses is irrelevant. The formula for the associative property is:

$$(a + b) + c = a + (b + c)$$

Here are some examples of the associative property at work:

$$30 = (6 + 14) + 10 = 6 + (14 + 10) = 30$$

$$35 = 8 + (2 + 25) = (8 + 2) + 25 = 35$$

Subtraction

Subtraction is taking away one number from another, so their quantities are reduced. The sign designating a subtraction operation is the – symbol, and the result is called the **difference**. For example:

$$9 - 6 = 3$$

The number *6* detracts from the number *9* to reach the difference *3*.

Unlike addition, subtraction follows neither the commutative nor associative properties. The order and grouping in subtraction impact the result.

$$15 = 22 - 7 \neq 7 - 22 = -15$$

$$3 = (10 - 5) - 2 \neq 10 - (5 - 2) = 7$$

When working through subtraction problems involving larger numbers, it's necessary to regroup the numbers. Let's work through a practice problem using regrouping:

$$\begin{array}{r} 325 \\ -\ 77 \\ \hline \end{array}$$

Here, it is clear that the ones and tens columns for 77 are greater than the ones and tens columns for 325. To subtract this number, borrow from the tens and hundreds columns. When borrowing from a column, subtracting 1 from the lender column will add 10 to the borrower column:

$$\begin{array}{rrr} 3-1 & 10+2-1 & 10+5 \\ - & 7 & 7 \\ \hline \end{array} = \begin{array}{rrr} 2 & 11 & 15 \\ - & 7 & 7 \\ \hline 2 & 4 & 8 \end{array}$$

After ensuring that each digit in the top row is greater than the digit in the corresponding bottom row, subtraction can proceed as normal, and the answer is found to be 248.

Multiplication

Multiplication involves adding together multiple copies of a number. It is indicated by an $\times$ symbol or a number immediately outside of a parenthesis. For example:

$$5(8-2)$$

The two numbers being multiplied together are called **factors**, and their result is called a **product**. For example:

$$9 \times 6 = 54$$

This can be shown alternatively by expansion of either the 9 or the 6:

$$9 \times 6 = 9 + 9 + 9 + 9 + 9 + 9 = 54$$

$$9 \times 6 = 6 + 6 + 6 + 6 + 6 + 6 + 6 + 6 + 6 = 54$$

Like addition, multiplication holds the commutative and associative properties:

$$115 = 23 \times 5 = 5 \times 23 = 115$$

$$84 = 3 \times (7 \times 4) = (3 \times 7) \times 4 = 84$$

Multiplication also follows the distributive property, which allows the multiplication to be distributed through parentheses. The formula for distribution is $a \times (b + c) = ab + ac$. This is clear after the examples:

$$45 = 5 \times 9 = 5(3 + 6)$$

$$(5 \times 3) + (5 \times 6) = 15 + 30 = 45$$

$$20 = 4 \times 5 = 4(10 - 5)$$

$$(4 \times 10) - (4 \times 5) = 40 - 20 = 20$$

Multiplication becomes slightly more complicated when multiplying numbers with decimals. The easiest way to answer these problems is to ignore the decimals and multiply as if they were whole numbers. After multiplying the factors, place a decimal in the product. The placement of the decimal is determined by taking the cumulative number of decimal places in the factors.

For example:

$$\begin{array}{r} 0.7 \\ \times\ 3 \\ \hline 2.1 \end{array} \qquad \begin{array}{r} 2.6 \\ \times\ 4.2 \\ \hline 10.92 \end{array} \qquad \begin{array}{r} 1.5 \\ \times\ 6.4 \\ \hline 9.60 \end{array}$$

Let's tackle the first example. First, ignore the decimal and multiply the numbers as though they were whole numbers to arrive at a product: 21. Second, count the number of digits that follow a decimal (one). Finally, move the decimal place that many positions to the left, as the factors have only one decimal place. The second example works the same way, except that there are two total decimal places in the factors, so the product's decimal is moved two places over. In the third example, the decimal should be moved over two digits, but the digit zero is no longer needed, so it is erased, and the final answer is 9.6.

Division

Division and multiplication are inverses of each other in the same way that addition and subtraction are opposites. The signs designating a division operation are the ÷ and / symbols. In division, the second number divides into the first.

The number before the division sign is called the **dividend** or, if expressed as a fraction, the **numerator**. For example, in $a \div b$, a is the dividend, while in $\frac{a}{b}$, a is the numerator.

The number after the division sign is called the **divisor** or, if expressed as a fraction, the **denominator**. For example, in $a \div b$, b is the divisor, while in $\frac{a}{b}$, b is the denominator.

Like subtraction, division doesn't follow the commutative property, as it matters which number comes before the division sign, and division doesn't follow the associative or distributive properties for the same reason.

For example:

$$\frac{3}{2} = 9 \div 6 \neq 6 \div 9 = \frac{2}{3}$$

$$2 = 10 \div 5 = (30 \div 3) \div 5 \neq 30 \div (3 \div 5) = 30 \div \frac{3}{5} = 50$$

$$25 = 20 + 5 = (40 \div 2) + (40 \div 8) \neq 40 \div (2 + 8) = 40 \div 10 = 4$$

If a divisor doesn't divide into a dividend an integer number of times, whatever is left over is termed the **remainder**. The remainder can be further divided out into decimal form by using long division; however, this doesn't always give a quotient with a finite number of decimal places, so the remainder can also be expressed as a fraction over the original divisor.

Division with decimals is similar to multiplication with decimals in that when dividing a decimal by a whole number, ignore the decimal and divide as if it were a whole number.

Upon finding the answer, or quotient, place the decimal at the decimal place equal to that in the dividend.

$$15.75 \div 3 = 5.25$$

When the divisor is a decimal number, multiply both the divisor and dividend by 10. Repeat this until the divisor is a whole number, then complete the division operation as described above.

$$17.5 \div 2.5 = 175 \div 25 = 7$$

Problem Situations for Operations

Addition and subtraction are inverse operations. Adding a number and then subtracting the same number will cancel each other out, resulting in the original number, and vice versa. For example:

$$8 + 7 - 7 = 8$$

and

$$137 - 100 + 100 = 137$$

Similarly, multiplication and division are inverse operations. Therefore, multiplying by a number and then dividing by the same number results in the original number, and vice versa. For example:

$$8 \times 2 \div 2 = 8$$

and

$$12 \div 4 \times 4 = 12$$

Inverse operations are used to work backwards to solve problems. In the case that 7 and a number add to 18, the inverse operation of subtraction is used to find the unknown value ($18 - 7 = 11$). If a school's entire 4th grade was divided evenly into 3 classes each with 22 students, the inverse operation of multiplication is used to determine the total students in the grade ($22 \times 3 = 66$). Additional scenarios involving inverse operations are included in the tables below.

There are a variety of real-world situations in which one or more of the operators is used to solve a problem. The tables below display the most common scenarios.

Addition & Subtraction

	Unknown Result	**Unknown Change**	**Unknown Start**
Adding to	5 students were in class. 4 more students arrived. How many students are in class? $5 + 4 = ?$	8 students were in class. More students arrived late. There are now 18 students in class. How many students arrived late? $8 + ? = 18$ Solved by inverse operations $18 - 8 = ?$	Some students were in class early. 11 more students arrived. There are now 17 students in class. How many students were in class early? $? + 11 = 17$ Solved by inverse operations $17 - 11 = ?$
Taking from	15 students were in class. 5 students left class. How many students are in class now? $15 - 5 = ?$	12 students were in class. Some students left class. There are now 8 students in class. How many students left class? $12 - ? = 8$ Solved by inverse operations $8 + ? = 12 \rightarrow 12 - 8 = ?$	Some students were in class. 3 students left class. Then there were 13 students in class. How many students were in class before? $? - 3 = 13$ Solved by inverse operations $13 + 3 = ?$

	Unknown Total	**Unknown Addends (Both)**	**Unknown Addends (One)**
Putting together/ taking apart	The homework assignment is 10 addition problems and 8 subtraction problems. How many problems are in the homework assignment? $10 + 8 = ?$	Bobby has $9. How much can Bobby spend on candy and how much can Bobby spend on toys? $9 = ? + ?$	Bobby has 12 pairs of pants. 5 pairs of pants are shorts, and the rest are long. How many pairs of long pants does he have? $12 = 5 + ?$ Solved by inverse operations $12 - 5 = ?$

	Unknown Difference	Unknown Larger Value	Unknown Smaller Value
Comparing	Bobby has 5 toys. Tommy has 8 toys. How many more toys does Tommy have than Bobby? $5 + ? = 8$ Solved by inverse operations $8 - 5 = ?$ Bobby has $6. Tommy has $10. How many fewer dollars does Bobby have than Tommy? $10 - 6 = ?$	Tommy has 2 more toys than Bobby. Bobby has 4 toys. How many toys does Tommy have? $2 + 4 = ?$ Bobby has 3 fewer dollars than Tommy. Bobby has $8. How many dollars does Tommy have? $? - 3 = 8$ Solved by inverse operations $8 + 3 = ?$	Tommy has 6 more toys than Bobby. Tommy has 10 toys. How many toys does Bobby have? $? + 6 = 10$ Solved by inverse operations $10 - 6 = ?$ Bobby has $5 less than Tommy. Tommy has $9. How many dollars does Bobby have? $9 - 5 = ?$

Multiplication and Division

	Unknown Product	Unknown Group Size	Unknown Number of Groups
Equal groups	There are 5 students, and each student has 4 pieces of candy. How many pieces of candy are there in all? $5 \times 4 = ?$	14 pieces of candy are shared equally by 7 students. How many pieces of candy does each student have? $7 \times ? = 14$ Solved by inverse operations $14 \div 7 = ?$	If 18 pieces of candy are to be given out 3 to each student, how many students will get candy? $? \times 3 = 18$ Solved by inverse operations $18 \div 3 = ?$

	Unknown Product	Unknown Factor	Unknown Factor
Arrays	There are 5 rows of students with 3 students in each row. How many students are there? $5 \times 3 = ?$	If 16 students are arranged into 4 equal rows, how many students will be in each row? $4 \times ? = 16$ Solved by inverse operations $16 \div 4 = ?$	If 24 students are arranged into an array with 6 columns, how many rows are there? $? \times 6 = 24$ Solved by inverse operations $24 \div 6 = ?$

	Larger Unknown	Smaller Unknown	Multiplier Unknown
Comparing	A small popcorn costs $1.50. A large popcorn costs 3 times as much as a small popcorn. How much does a large popcorn cost? $1.50 \times 3 = ?$	A large soda costs $6 and that is 2 times as much as a small soda costs. How much does a small soda cost? $2 \times ? = 6$ Solved by inverse operations $6 \div 2 = ?$	A large pretzel costs $3 and a small pretzel costs $2. How many times as much does the large pretzel cost as the small pretzel? $? \times 2 = 3$ Solved by inverse operations $3 \div 2 = ?$

Order of Operations

When working with complicated expressions, parentheses are used to indicate in which order to perform operations. However, to avoid having too many parentheses in an expression, here are some basic rules concerning the proper order to perform operations when not otherwise specified.

1. Parentheses: always perform operations inside parentheses first, regardless of what those operations are
2. Exponents
3. Multiplication and Division
4. Addition and Subtraction

For #3 & #4, work these from left to right. So, if there a subtraction problem and then an addition problem, the subtraction problem will be worked first.

Note that multiplication and division are performed from left to right as they appear in the expression or equation. Addition and subtraction also are performed from left to right as they appear.

As an aid to memorizing this, some students like to use the mnemonic **PEMDAS**. Furthermore, this acronym can be associated with a mnemonic phrase such as "Pirates Eat Many Donuts At Sea."

Evaluate the following two problems to understand the Order of Operations:

1) $4 + (3 \times 2)^2 \div 4$

First, solve the operation within the parentheses: $4 + 6^2 \div 4$.
Second, solve the exponent: $4 + 36 \div 4$.
Third, solve the division operation: $4 + 9$.
Fourth, finish the operation with addition for the answer, 13.

2) $2 \times (6 + 3) \div (2 + 1)^2$

$2 \times 9 \div (3)^2$
$2 \times 9 \div 9$
$18 \div 9$
2

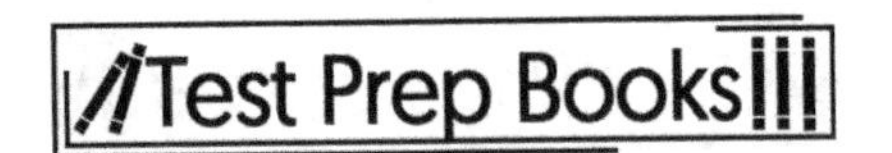

Properties of Operations

Properties of operations exist that make calculations easier and solve problems for missing values. The following table summarizes commonly used properties of real numbers.

Property	Addition	Multiplication
Commutative	$a + b = b + a$	$a \times b = b \times a$
Associative	$(a + b) + c = a + (b + c)$	$(a \times b) \times c = a \times (b \times c)$
Identity	$a + 0 = a; 0 + a = a$	$a \times 1 = a; 1 \times a = a$
Inverse	$a + (-a) = 0$	$a \times \frac{1}{a} = 1; a \neq 0$
Distributive	$a(b + c) = ab + ac$	

The **cumulative property** of addition states that the order in which numbers are added does not change the sum. Similarly, the **commutative property** of multiplication states that the order in which numbers are multiplied does not change the product. The **associative property** of addition and multiplication state that the grouping of numbers being added or multiplied does not change the sum or product, respectively. The commutative and associative properties are useful for performing calculations. For example:

$$(47 + 25) + 3$$

is equivalent to:

$$(47 + 3) + 25$$

which is easier to calculate.

The **identity property** of addition states that adding zero to any number does not change its value. The identity property of multiplication states that multiplying a number by 1 does not change its value. The **inverse property** of addition states that the sum of a number and its opposite equals zero. Opposites are numbers that are the same with different signs (ex. 5 and -5; $-\frac{1}{2}$ and $\frac{1}{2}$). The inverse property of multiplication states that the product of a number (other than 0) and its reciprocal equals 1. Reciprocal numbers have numerators and denominators that are inverted (ex. $\frac{2}{5}$ and $\frac{5}{2}$). Inverse properties are useful for canceling quantities to find missing values (see algebra content). For example:

$$a + 7 = 12$$

is solved by adding the inverse of 7 (which is -7) to both sides in order to isolate *a*.

The **distributive property** states that multiplying a sum (or difference) by a number produces the same result as multiplying each value in the sum (or difference) by the number and adding (or subtracting) the products. Consider the following scenario: You are buying three tickets for a baseball game. Each ticket costs $18. You are also charged a fee of $2 per ticket for purchasing the tickets online. The cost is calculated:

$$3 \times 18 + 3 \times 2$$

Using the distributive property, the cost can also be calculated $3(18 + 2)$.

Comparing, Classifying, and Ordering Real Numbers

A **rational number** is any number that can be written as a fraction or ratio. Within the set of rational numbers, several subsets exist that are referenced throughout the mathematics topics. **Counting numbers** are the first numbers learned as a child. Counting numbers consist of 1,2,3,4, and so on. **Whole numbers** include all counting numbers and zero (0,1,2,3, 4,...). **Integers** include counting numbers, their opposites, and zero (...,-3,-2,-1,0,1,2,3,...). **Rational numbers** are inclusive of integers, fractions, and decimals that terminate, or end (1.7, 0.04213) or repeat ($0.136\bar{5}$).

A **number line** typically consists of integers (...3,2,1,0,-1,-2,-3...), and is used to visually represent the value of a rational number. Each rational number has a distinct position on the line determined by comparing its value with the displayed values on the line. For example, if plotting -1.5 on the number line below, it is necessary to recognize that the value of -1.5 is .5 less than -1 and .5 greater than -2. Therefore, -1.5 is plotted halfway between -1 and -2.

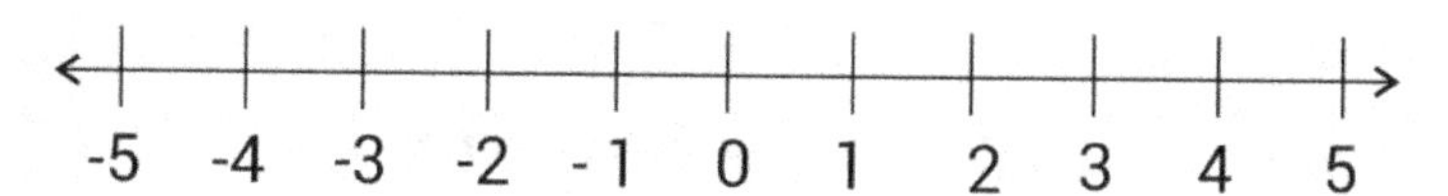

The number system that is used consists of only ten different digits or characters. However, this system is used to represent an infinite number of values. The **place value system** makes this infinite number of values possible. The position in which a digit is written corresponds to a given value. Starting from the decimal point (which is implied, if not physically present), each subsequent place value to the left represents a value greater than the one before it. Conversely, starting from the decimal point, each subsequent place value to the right represents a value less than the one before it.

The names for the place values to the left of the decimal point are as follows:

...	Billions	Hundred-Millions	Ten-Millions	Millions	Hundred-Thousands	Ten-Thousands	Thousands	Hundreds	Tens	Ones

*Note that this table can be extended infinitely further to the left.

The names for the place values to the right of the decimal point are as follows:

Decimal Point (.)	Tenths	Hundredths	Thousandths	Ten-Thousandths	...

*Note that this table can be extended infinitely further to the right.

When given a multi-digit number, the value of each digit depends on its place value. Consider the number 682,174.953. Referring to the chart above, it can be determined that the digit 8 is in the ten-thousands place. It is in the fifth place to the left of the decimal point. Its value is 8 ten-thousands or 80,000. The digit 5 is two places to the right of the decimal point. Therefore, the digit 5 is in the hundredths place. Its value is 5 hundredths or $\frac{5}{100}$ (equivalent to .05).

In accordance with the **base-10 system**, the value of a digit increases by a factor of ten each place it moves to the left. For example, consider the number 7. Moving the digit one place to the left (70), increases its value by a factor of 10:

$$(7 \times 10 = 70)$$

Moving the digit two places to the left (700) increases its value by a factor of 10 twice:

$$(7 \times 10 \times 10 = 700)$$

Moving the digit three places to the left (7,000) increases its value by a factor of 10 three times $(7 \times 10 \times 10 \times 10 = 7{,}000)$, and so on.

Conversely, the value of a digit decreases by a factor of ten each place it moves to the right. (Note that multiplying by $\frac{1}{10}$ is equivalent to dividing by 10). For example, consider the number 40. Moving the digit one place to the right (4) decreases its value by a factor of 10:

$$(40 \div 10 = 4)$$

Moving the digit two places to the right (0.4), decreases its value by a factor of 10 twice:

$$(40 \div 10 \div 10 = 0.4)$$

or

$$(40 \times \frac{1}{10} \times \frac{1}{10} = 0.4)$$

Moving the digit three places to the right (0.04) decreases its value by a factor of 10 three times:

$$(40 \div 10 \div 10 \div 10 = 0.04)$$

or

$$(40 \times \frac{1}{10} \times \frac{1}{10} \times \frac{1}{10} = 0.04)$$

and so on.

Ordering Numbers

A common question type asks to order rational numbers from least to greatest or greatest to least. The numbers will come in a variety of formats, including decimals, percentages, roots, fractions, and whole numbers. These questions test for knowledge of different types of numbers and the ability to determine their respective values.

Before discussing ordering all numbers, let's start with decimals.

To compare decimals and order them by their value, utilize a method similar to that of ordering large numbers.

The main difference is where the comparison will start. Assuming that any numbers to left of the decimal point are equal, the next numbers to be compared are those immediately to the right of the decimal point. If those are equal, then move on to compare the values in the next decimal place to the right.

For example:

Which number is greater, 12.35 or 12.38?

Check that the values to the left of the decimal point are equal:

$$12 = 12$$

Next, compare the values of the decimal place to the right of the decimal:

$$12.3 = 12.3$$

Those are also equal in value.

Finally, compare the value of the numbers in the next decimal place to the right on both numbers:

12.3**5** and 12.3**8**

Here the 5 is less than the 8, so the final way to express this inequality is:

$$12.35 < 12.38$$

Comparing decimals is regularly exemplified with money because the "cents" portion of money ends in the hundredths place. When paying for gasoline or meals in restaurants, and even in bank accounts, if enough errors are made when calculating numbers to the hundredths place, they can add up to dollars and larger amounts of money over time.

Now that decimal ordering has been explained, let's expand and consider all real numbers. Whether the question asks to order the numbers from greatest to least or least to greatest, the crux of the question is the same—convert the numbers into a common format. Generally, it's easiest to write the numbers as whole numbers and decimals so they can be placed on a number line. Follow these examples to understand this strategy.

1) Order the following rational numbers from greatest to least:

$$\sqrt{36}, 0.65, 78\%, \frac{3}{4}, 7, 90\%, \frac{5}{2}$$

Of the seven numbers, the whole number (7) and decimal (0.65) are already in an accessible form, so concentrate on the other five.

First, the square root of 36 equals 6. If the test asks for the root of a non-perfect root, determine which two whole numbers the root lies between. Next, convert the percentages to decimals. A percentage means "per hundred," so this conversion requires moving the decimal point two places to the left, leaving 0.78 and 0.9. Lastly, evaluate the fractions:

$$\frac{3}{4} = \frac{75}{100} = 0.75\,; \frac{5}{2} = 2\frac{1}{2} = 2.5$$

Now, the only step left is to list the numbers in the request order:

$$7, \sqrt{36}, \frac{5}{2}, 90\%, 78\%, \frac{3}{4}, 0.65$$

2) Order the following rational numbers from least to greatest:

$$2.5, \sqrt{9}, -10.5, 0.853, 175\%, \sqrt{4}, \frac{4}{5}$$

$$\sqrt{9} = 3$$

$$175\% = 1.75$$

$$\sqrt{4} = 2$$

$$\frac{4}{5} = 0.8$$

From least to greatest, the answer is:

$$-10.5, \frac{4}{5}, 0.853, 175\%, \sqrt{4}, 2.5, \sqrt{9}$$

Basic Concepts of Number Theory

Prime and Composite Numbers

Whole numbers are classified as either prime or composite. A **prime number** can only be divided evenly by itself and one. For example, the number 11 can only be divided evenly by 11 and one; therefore, 11 is a prime number. A helpful way to visualize a prime number is to use concrete objects and try to divide them into equal piles. If dividing 11 coins, the only way to divide them into equal piles is to create 1 pile of 11 coins or to create 11 piles of 1 coin each. Other examples of prime numbers include 2, 3, 5, 7, 13, 17, and 19.

A **composite number** is any whole number that is not a prime number. A composite number is a number that can be divided evenly by one or more numbers other than itself and one. For example, the number 6 can be divided evenly by 2 and 3. Therefore, 6 is a composite number. If dividing 6 coins into equal piles, the possibilities are 1 pile of 6 coins, 2 piles of 3 coins, 3 piles of 2 coins, or 6 piles of 1 coin. Other examples of composite numbers include 4, 8, 9, 10, 12, 14, 15, 16, 18, and 20.

To determine if a number is a prime or composite number, the number is divided by every whole number greater than one and less than its own value. If it divides evenly by any of these numbers, then the number is composite. If it does not divide evenly by any of these numbers, then the number is prime. For example, when attempting to divide the number 5 by 2, 3, and 4, none of these numbers divide evenly. Therefore, 5 must be a prime number.

Odd and Even Numbers

Even numbers are all divisible by the number 2. **Odd numbers** are not divisible by 2, and an odd quantity of items cannot be paired up into groups of 2 without having 1 item leftover. Examples of even numbers are 2, 4, 6, 20, 100, 242, etc. Examples of odd numbers are 1, 3, 5, 27, 99, 333, etc.

Factors, Multiples, and Divisibility

The **Fundamental Theorem of Arithmetic** states that any integer greater than 1 is either a prime number or can be written as a unique product of prime numbers. Factors can be used to find the combination of numbers to multiply to produce an integer that is not prime. The factors of a number are all integers that can be multiplied by another integer to produce the given number. For example, 2 is multiplied by 3 to produce 6. Therefore, 2 and 3 are both factors of 6. Similarly:

$$1 \times 6 = 6$$

and

$$2 \times 3 = 6$$

so 1, 2, 3, and 6 are all factors of 6. Another way to explain a factor is to say that a given number divides evenly by each of its factors to produce an integer. For example, 6 does not divide evenly by 5. Therefore, 5 is not a factor of 6.

Essentially, **factors** are the numbers multiplied to achieve a product. Thus, every product in a multiplication equation has, at minimum, two factors. Of course, some products will have more than two factors. For the sake of most discussions, assume that factors are positive integers.

To find a number's factors, start with 1 and the number itself. Then divide the number by 2, 3, 4, and so on, seeing if any divisors can divide the number without a remainder, keeping a list of those that do. Stop upon reaching either the number itself or another factor.

Let's find the factors of 45. Start with 1 and 45. Then try to divide 45 by 2, which fails. Now divide 45 by 3. The answer is 15, so 3 and 15 are now factors. Dividing by 4 doesn't work and dividing by 5 leaves 9. Lastly, dividing 45 by 6, 7, and 8 all don't work. The next integer to try is 9, but this is already known to be a factor, so the factorization is complete. The factors of 45 are 1, 3, 5, 9, 15 and 45.

A **common factor** is a factor shared by two numbers. Let's take 45 and 30 and find the common factors:

The factors of 45 are: 1, 3, 5, 9, 15, and 45.
The factors of 30 are: 1, 2, 3, 5, 6, 10, 15, and 30.
The common factors are 1, 3, 5, and 15.

The **greatest common factor** is the largest number among the shared, common factors. From the factors of 45 and 30, the common factors are 3, 5, and 15. Thus, 15 is the greatest common factor, as it's the largest number.

Multiples of a given number are found by taking that number and multiplying it by any other whole number. For example, 3 is a factor of 6, 9, and 12. Therefore, 6, 9, and 12 are multiples of 3. The multiples of any number are an infinite list. For example, the multiples of 5 are 5, 10, 15, 20, and so on. This list continues without end. A list of multiples is used in finding the **least common multiple**, or **LCM,** for fractions when a common denominator is needed. The denominators are written down and their multiples listed until a common number is found in both lists. This common number is the LCM.

If two numbers share no factors besides 1 in common, then their least common multiple will be simply their product. If two numbers have common factors, then their least common multiple will be their product divided by their greatest common factor. This can be visualized by the formula:

$$LCM = \frac{x \times y}{GCF}$$

here x and y are some integers, and LCM and GCF are their least common multiple and greatest common factor, respectively.

Prime factorization breaks down each factor of a whole number until only prime numbers remain. All composite numbers can be factored into prime numbers. For example, the prime factors of 12 are 2, 2, and 3 ($2 \times 2 \times 3 = 12$). To produce the prime factors of a number, the number is factored, and any composite numbers are continuously factored until the result is the product of prime factors only. A factor tree, such as the one below, is helpful when exploring this concept.

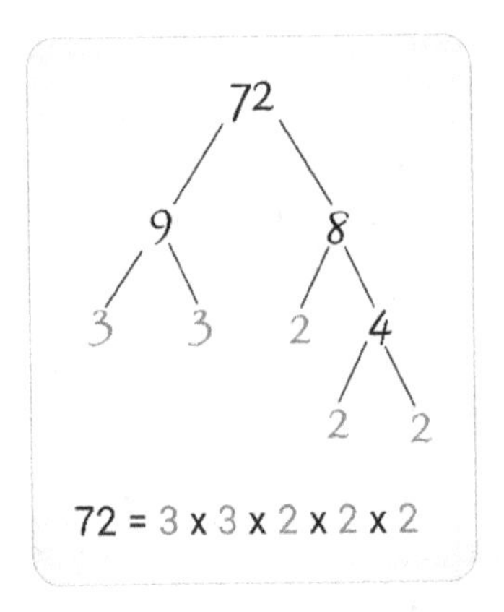

Let's break 129 down into its prime factors. First, the factors are 3 and 43. Both 3 and 43 are prime numbers, so we're done. But if 43 was not a prime number, then it would also need to be factorized until all of the factors are expressed as prime numbers.

Fractions

A **fraction** is an equation that represents a part of a whole but can also be used to present ratios or division problems. An example of a fraction is $\frac{x}{y}$. In this example, x is called the **numerator,** while y is the **denominator**. The numerator represents the number of parts, and the denominator is the total number of parts. They are separated by a line or slash, known as a fraction bar. In simple fractions, the numerator and denominator can be nearly any integer. However, the denominator of a fraction can never be zero because dividing by zero is a function that is undefined.

Imagine that an apple pie has been baked for a holiday party, and the full pie has eight slices. After the party, there are five slices left. How could the amount of the pie that remains be expressed as a fraction? The numerator is 5 since there are 5 pieces left, and the denominator is 8 since there were eight total slices in the whole pie. Thus, expressed as a fraction, the leftover pie totals $\frac{5}{8}$ of the original amount.

Fractions come in three different varieties: proper fractions, improper fractions, and mixed numbers. **Proper fractions** have a numerator less than the denominator, such as $\frac{3}{8}$, but **improper fractions** have a numerator greater than the denominator, such as $\frac{15}{8}$. **Mixed numbers** combine a whole number with a proper fraction, such as $3\frac{1}{2}$. Any mixed number can be written as an improper fraction by multiplying the integer by the denominator, adding the product to the value of the numerator, and dividing the sum by the original denominator. For example:

$$3\frac{1}{2} = \frac{3 \times 2 + 1}{2} = \frac{7}{2}$$

Whole numbers can also be converted into fractions by placing the whole number as the numerator and making the denominator 1. For example:

$$3 = \frac{3}{1}$$

One of the most fundamental concepts of fractions is their ability to be manipulated by multiplication or division. This is possible since $\frac{n}{n} = 1$ for any non-zero integer. As a result, multiplying or dividing by $\frac{n}{n}$ will not alter the original fraction since any number multiplied or divided by 1 doesn't change the value of that number. Fractions of the same value are known as equivalent fractions.

For example, $\frac{2}{4}, \frac{4}{8}, \frac{50}{100}$, and $\frac{75}{150}$ are equivalent, as they all equal $\frac{1}{2}$.

Although many equivalent fractions exist, they are easier to compare and interpret when reduced or simplified. The numerator and denominator of a simple fraction will have no factors in common other than 1. When reducing or simplifying fractions, divide the numerator and denominator by the greatest common factor. A simple strategy is to divide the numerator and denominator by low numbers, like 2, 3, or 5, until arriving at a simple fraction, but the same thing could be achieved by determining the greatest common factor for both the numerator and denominator and dividing each by it. Using the first method is preferable when both the numerator and denominator are even, end in 5, or are obviously a multiple of another number. However, if no numbers seem to work, it will be necessary to factor the numerator and denominator to find the GCF. Let's look at examples:

1) Simplify the fraction $\frac{6}{8}$:

Dividing the numerator and denominator by 2 results in $\frac{3}{4}$, which is a simple fraction.

2) Simplify the fraction $\frac{12}{36}$:

Dividing the numerator and denominator by 2 leaves $\frac{6}{18}$. This isn't a simple fraction, as both the numerator and denominator have factors in common. Dividing each by 3 results in $\frac{2}{6}$, but this can be further simplified by dividing by 2 to get $\frac{1}{3}$. This is the simplest fraction, as the numerator is 1. In cases like this, multiple division operations can be avoided by determining the greatest common factor between the numerator and denominator.

3) Simplify the fraction $\frac{18}{54}$ by dividing by the greatest common factor:

First, determine the factors for the numerator and denominator. The factors of 18 are 1, 2, 3, 6, 9, and 18. The factors of 54 are 1, 2, 3, 6, 9, 18, 27, and 54. Thus, the greatest common factor is 18. Dividing $\frac{18}{54}$ by 18 leaves $\frac{1}{3}$, which is the simplest fraction. This method takes slightly more work, but it definitively arrives at the simplest fraction.

Of the four basic operations that can be performed on fractions, the one that involves the least amount of work is multiplication. To multiply two fractions, simply multiply the numerators together, multiply the denominators together, and place the products of each as a fraction. Whole numbers and mixed numbers can also be expressed as a fraction, as described above, to multiply with a fraction. Let's work through a couple of examples.

1) $\frac{2}{5} \times \frac{3}{4} = \frac{6}{20} = \frac{3}{10}$

2) $\frac{4}{9} \times \frac{7}{11} = \frac{28}{99}$

Dividing fractions is similar to multiplication with one key difference. To divide fractions, flip the numerator and denominator of the second fraction, and then proceed as if it were a multiplication problem:

1) $\frac{7}{8} \div \frac{4}{5} = \frac{7}{8} \times \frac{5}{4} = \frac{35}{32}$

2) $\frac{5}{9} \div \frac{1}{3} = \frac{5}{9} \times \frac{3}{1} = \frac{15}{9} = \frac{5}{3}$

Addition and subtraction require more steps than multiplication and division, as these operations require the fractions to have the same denominator, also called a common denominator. It is always possible to find a common denominator by multiplying the denominators. However, when the denominators are large numbers, this method is unwieldy, especially if the answer must be provided in its simplest form. Thus, it's beneficial to find the least common denominator of the fractions—the least common denominator is incidentally also the least common multiple.

Once equivalent fractions have been found with common denominators, simply add or subtract the numerators to arrive at the answer:

1) $\frac{1}{2} + \frac{3}{4} = \frac{2}{4} + \frac{3}{4} = \frac{5}{4}$

2) $\frac{3}{12} + \frac{11}{20} = \frac{15}{60} + \frac{33}{60} = \frac{48}{60} = \frac{4}{5}$

3) $\frac{7}{9} - \frac{4}{15} = \frac{35}{45} - \frac{12}{45} = \frac{23}{45}$

4) $\frac{5}{6} - \frac{7}{18} = \frac{15}{18} - \frac{7}{18} = \frac{8}{18} = \frac{4}{9}$

Decimals

The **decimal system** is a way of writing out numbers that uses ten different numerals: 0, 1, 2, 3, 4, 5, 6, 7, 8, and 9. This is also called a "base ten" or "base 10" system. Other bases are also used. For example, computers work with a base of 2. This means they only use the numerals 0 and 1.

The **decimal place** denotes how far to the right of the decimal point a numeral is. The first digit to the right of the decimal point is in the *tenths* place. The next is the **hundredths**. The third is the **thousandths**.

So, 3.142 has a 1 in the tenths place, a 4 in the hundredths place, and a 2 in the thousandths place.

The **decimal point** is a period used to separate the **ones** place from the **tenths** place when writing out a number as a decimal.

A **decimal number** is a number written out with a decimal point instead of as a fraction, for example, 1.25 instead of $\frac{5}{4}$. Depending on the situation, it can sometimes be easier to work with fractions and sometimes easier to work with decimal numbers.

A decimal number is **terminating** if it stops at some point. It is called **repeating** if it never stops but repeats a pattern over and over. It is important to note that every rational number can be written as a terminating decimal or as a repeating decimal.

Addition with Decimals

To add decimal numbers, each number needs to be lined up by the decimal point in vertical columns. For each number being added, the zeros to the right of the last number need to be filled in so that each of the numbers has the same number of places to the right of the decimal. Then, the columns can be added together. Here is an example of $2.45 + 1.3 + 8.891$ written in column form:

$$2.450$$

$$1.300$$

$$\underline{+\ 8.891}$$

Zeros have been added in the columns so that each number has the same number of places to the right of the decimal.

Added together, the correct answer is 12.641:

$$2.450$$

$$1.300$$

$$\underline{+\ 8.891}$$

$$12.641$$

Subtraction with Decimals

Subtracting decimal numbers is the same process as adding decimals. Here is $7.89 - 4.235$ written in column form:

$$7.890$$

$$-\ 4.235$$

$$3.655$$

A zero has been added in the column so that each number has the same number of places to the right of the decimal.

Multiplication with Decimals

The simplest way to multiply decimals is to calculate the product as if the decimals are not there, then count the number of decimal places in the original problem. Use that total to place the decimal the same number of places over in your answer, counting from right to left. For example, 0.5×1.25 can be rewritten and multiplied as:

$$5 \times 125$$

which equals 625. Then the decimal is added three places from the right for .625.

The final answer will have the same number of decimal *points* as the total number of decimal *places* in the problem. The first number has one decimal place, and the second number has two decimal places. Therefore, the final answer will contain three decimal places:

$$0.5 \times 1.25 = 0.625$$

Division with Decimals

Dividing a decimal by a whole number entails using long division first by ignoring the decimal point. Then, the decimal point is moved the number of places given in the problem.

For example:

$$6.8 \div 4$$

can be rewritten as:

$$68 \div 4$$

which is 17. There is one non-zero integer to the right of the decimal point, so the final solution would have one decimal place to the right of the solution. In this case, the solution is 1.7.

Dividing a decimal by another decimal requires changing the divisor to a whole number by moving its decimal point. The decimal place of the dividend should be moved by the same number of places as the divisor. Then, the problem is the same as dividing a decimal by a whole number.

For example, $5.72 \div 1.1$ has a divisor with one decimal point in the denominator. The expression can be rewritten as $57.2 \div 11$ by moving each number one decimal place to the right to eliminate the decimal. The long division can be completed as $572 \div 11$ with a result of 52. Since there is one non-zero integer to the right of the decimal point in the problem, the final solution is 5.2.

In another example, $8 \div 0.16$ has a divisor with two decimal points in the denominator. The expression can be rewritten as $800 \div 16$ by moving each number two decimal places to the right to eliminate the decimal in the divisor. The long division can be completed with a result of 50.

Percentages

Think of **percentages** as fractions with a denominator of 100. In fact, percentage means "per hundred." Problems often require converting numbers from percentages, fractions, and decimals.

The basic percent equation is the following:

$$\frac{is}{of} = \frac{\%}{100}$$

The placement of numbers in the equation depends on what the question asks.

Example 1: Find 40% of 80.

Basically, the problem is asking, "What is 40% of 80?" The 40% is the percent, and 80 is the number to find the percent "of." The equation is:

$$\frac{x}{80} = \frac{40}{100}$$

Solving the equation by cross-multiplication, the problem becomes $100x = 80(40)$. Solving for x gives the answer: $x = 32$.

Example 2: What percent of 100 is 20?

The 20 fills in the "is" portion, while 100 fills in the "of." The question asks for the percent, so that will be x, the unknown. The following equation is set up:

$$\frac{20}{100} = \frac{x}{100}$$

Cross-multiplying yields the equation $100x = 20(100)$. Solving for x gives the answer of 20%.

Example 3: 30% of what number is 30?

The following equation uses the clues and numbers in the problem:

$$\frac{30}{x} = \frac{30}{100}$$

Cross-multiplying results in the equation $30(100) = 30x$. Solving for x gives the answer $x = 100$.

Converting Fractions, Decimals, and Percentages

Decimals and Percentages

Since a percentage is based on "per hundred," decimals and percentages can be converted by multiplying or dividing by 100. Practically speaking, this always amounts to moving the decimal point two places to the right or left, depending on the conversion. To convert a percentage to a decimal, move the decimal point two places to the left and remove the % sign. To convert a decimal to a percentage, move the decimal point two places to the right and add a "%" sign. Here are some examples:

$$65\% = 0.65$$
$$0.33 = 33\%$$
$$0.215 = 21.5\%$$
$$99.99\% = 0.9999$$
$$500\% = 5.00$$
$$7.55 = 755\%$$

Fractions and Percentages

Remember that a percentage is a number per one hundred. So, a percentage can be converted to a fraction by making the number in the percentage the numerator and putting 100 as the denominator:

$$43\% = \frac{43}{100}$$

$$97\% = \frac{97}{100}$$

Note that the percent symbol (%) kind of looks like a 0, a 1, and another 0. So, think of a percentage like 54% as 54 over 100.

To convert a fraction to a percent, follow the same logic. If the fraction happens to have 100 in the denominator, you're in luck. Just take the numerator and add a percent symbol:

$$\frac{28}{100} = 28\%$$

Otherwise, divide the numerator by the denominator to get a decimal:

$$\frac{9}{12} = 0.75$$

Then convert the decimal to a percentage:

$$0.75 = 75\%$$

Another option is to make the denominator equal to 100. Be sure to multiply the numerator and the denominator by the same number. For example:

$$\frac{3}{20} \times \frac{5}{5} = \frac{15}{100}$$

$$\frac{15}{100} = 15\%$$

Changing Fractions to Decimals

To change a fraction into a decimal, divide the denominator into the numerator until there are no remainders. There may be repeating decimals, so rounding is often acceptable. A straight line above the repeating portion denotes that the decimal repeats.

Example: Express $\frac{4}{5}$ as a decimal.

Set up the division problem.

$$5\overline{)4}$$

5 does not go into 4, so place the decimal and add a zero.

$$5\overline{)4.0}$$

5 goes into 40 eight times. There is no remainder.

$$\begin{array}{r} 0.8 \\ 5\overline{)4.0} \\ -4.0 \\ \hline 0 \end{array}$$

The solution is 0.8.

Example: Express $33\frac{1}{3}$ as a decimal.

Since the whole portion of the number is known, set it aside to calculate the decimal from the fraction portion.

Set up the division problem.

$$3\overline{)1}$$

3 does not go into 1, so place the decimal and add zeros. 3 goes into 10 three times.

$$\begin{array}{r} 0.333 \\ 3\overline{)1.000} \end{array}$$

This will repeat with a remainder of 1, so placing a line over the 3 denotes the repetition.

$$\begin{array}{r} 0.333 \\ 3\overline{)1.000} \\ -9 \\ \hline 10 \\ -9 \\ \hline 10 \end{array}$$

The solution is $0.\overline{3}$

Changing Decimals to Fractions

To change decimals to fractions, place the decimal portion of the number, the numerator, over the respective place value, the denominator, then reduce, if possible.

Example: Express 0.25 as a fraction.

This is read as twenty-five hundredths, so put 25 over 100. Then reduce to find the solution.

$$\frac{25}{100} = \frac{1}{4}$$

Example: Express 0.455 as a fraction

This is read as four hundred fifty-five thousandths, so put 455 over 1000. Then reduce to find the solution.

$$\frac{455}{1000} = \frac{91}{200}$$

There are two types of problems that commonly involve percentages. The first is to calculate some percentage of a given quantity, where you convert the percentage to a decimal, and multiply the quantity by that decimal. Secondly, you are given a quantity and told it is a fixed percent of an unknown quantity. In this case, convert to a decimal, then divide the given quantity by that decimal.

Example: What is 30% of 760?

Convert the percent into a useable number. "Of" means to multiply.

$$30\% = 0.30$$

Set up the problem based on the givens and solve.

$$0.30 \times 760 = 228$$

Example: 8.4 is 20% of what number?

Convert the percent into a useable number.

$$20\% = 0.20$$

The given number is a percent of the answer needed, so divide the given number by this decimal rather than multiplying it.

$$\frac{8.4}{0.20} = 42$$

Representing Fractions, Decimals, and Percent Using Various Models

A **fraction** is a part of something that is whole. Items such as apples can be cut into parts to help visualize fractions. If an apple is cut into 2 equal parts, each part represents ½ of the apple. If each half is cut into two parts, the apple now is cut into quarters. Each piece now represents ¼ of the apple. In this example, each part is equal because they all have the same size. Geometric shapes, such as circles and squares, can also be utilized in the classroom to help visualize the idea of fractions.

For example, a circle can be drawn on the board and divided into 6 equal parts:

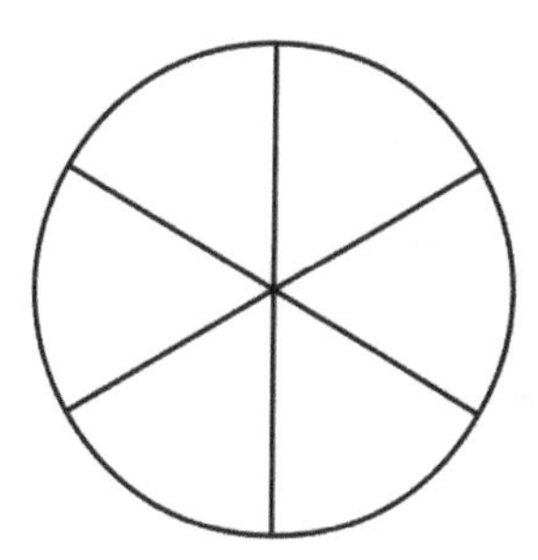

Shading can be used to represent parts of the circle that can be translated into fractions. The top of the fraction, the **numerator,** can represent how many segments are shaded. The bottom of the fraction, the **denominator,** can represent the number of segments that the circle is broken into. A pie is a good analogy to use in this example. If one piece of the circle is shaded, or one piece of pie is cut out, $^1/_6$ of the object is being referred to. An apple, a pie, or a circle can be utilized in order to compare simple fractions. For example, showing that $^1/_2$ is larger than $^1/_4$ and that $^1/_4$ is smaller than $^1/_3$ can be accomplished through shading. A **unit fraction** is a fraction in which the numerator is 1, and the denominator is a positive whole number. It represents one part of a whole—one piece of pie.

When representing fractions, it is important to remember the meaning of the word *fraction*. The number is a fraction, or portion, of a whole. The whole, or 1, is the number on the bottom part of the fraction, the denominator. For example, the fraction $\frac{2}{5}$ represents a number of portions (the numerator, 2) of how many it would take to make a whole (the denominator, 5). So, $\frac{2}{5}$ represents 2 portions out of the 5 it would take to make a whole.

This could also be represented with blocks, as follows:

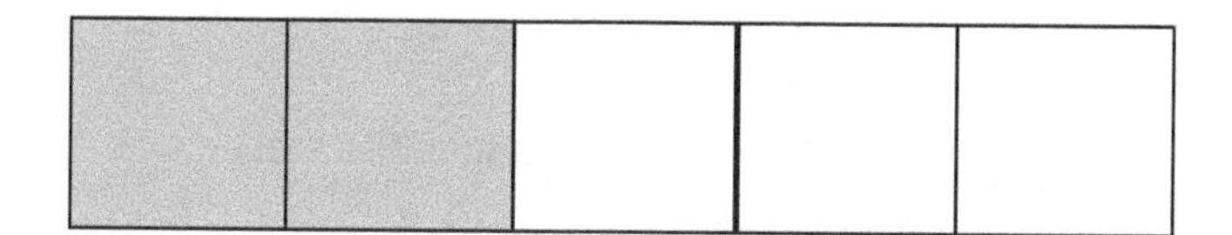

5 blocks = 1 whole, so only 2 of the 5 are shaded.

This method could also be used to represent fractions with a higher number in the numerator than in the denominator.

What does the fraction $\frac{6}{5}$ look like with the block method?

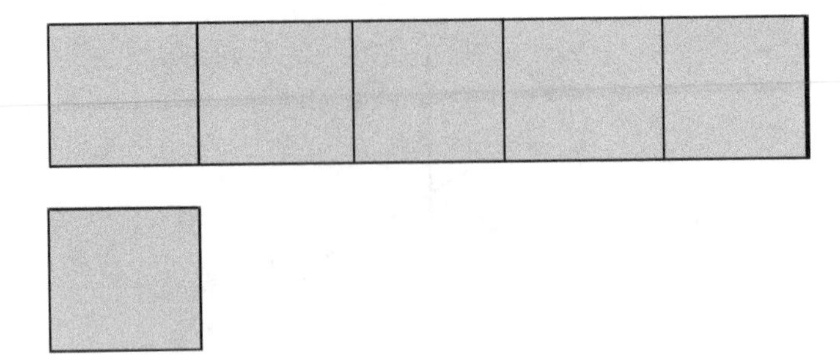

Fractions can be broken apart into sums of fractions with the same denominator. For example, the fraction $\frac{5}{6}$ can be decomposed into sums of fractions with all denominators equal to 6 and the numerators adding to 5. The fraction $\frac{5}{6}$ is decomposed as the following:

$$\frac{3}{6}+\frac{2}{6}$$

$$\frac{2}{6}+\frac{2}{6}+\frac{1}{6}$$

$$\frac{3}{6}+\frac{1}{6}+\frac{1}{6}$$

$$\frac{1}{6}+\frac{1}{6}+\frac{1}{6}+\frac{2}{6}$$

$$\frac{1}{6}+\frac{1}{6}+\frac{1}{6}+\frac{1}{6}+\frac{1}{6}$$

As mentioned, unit fraction is a fraction in which the numerator is 1. If decomposing a fraction into unit fractions, the sum will consist of a unit fraction added the number of times equal to the numerator. For example:

$$\frac{3}{4}=\frac{1}{4}+\frac{1}{4}+\frac{1}{4} \text{ (unit fractions } \frac{1}{4} \text{ added 3 times)}$$

Composing fractions is simply the opposite of decomposing. It is the process of adding fractions with the same denominators to produce a single fraction. For example:

$$\frac{3}{7}+\frac{2}{7}=\frac{5}{7}$$

and

$$\frac{1}{5}+\frac{1}{5}+\frac{1}{5}=\frac{3}{5}$$

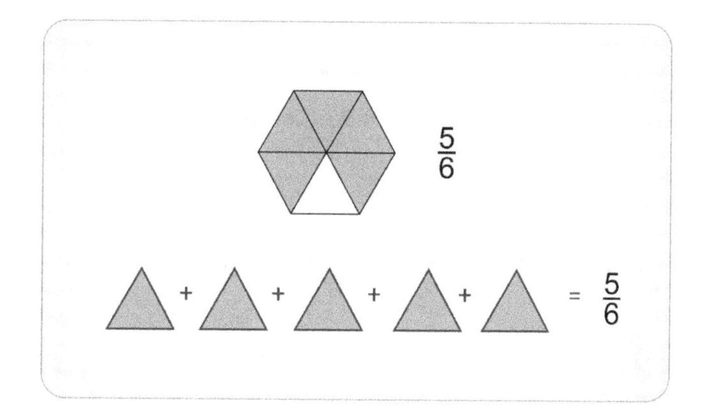

To represent fractions and decimals as distances beginning at zero on a number line, it's helpful to relate the fraction to a real-world application. For example, a charity walk covers $\frac{3}{10}$ of a mile. How could this distance be represented on a number line?

First, divide the number line into tenths, as follows:

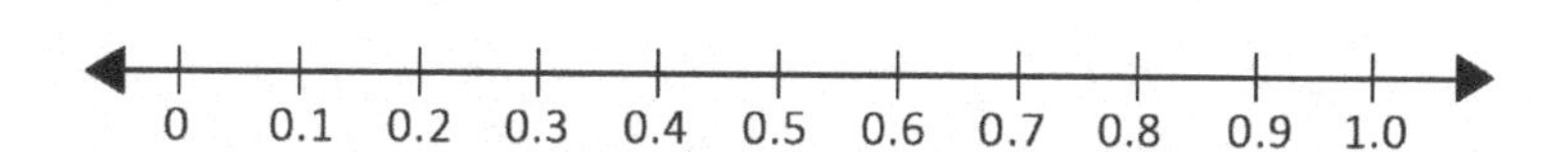

If each division on the number line represents one-tenth of one, or $\frac{1}{10}$, then representing the distance of the charity walk, $\frac{3}{10}$, would cover 3 of those divisions and look as follows:

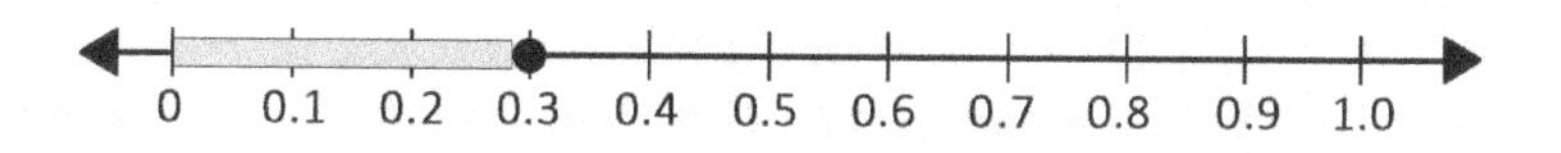

So, the fraction $\frac{3}{10}$ is represented by covering from 0 to 0.3 (or 3 sections) on the number line.

In general, to precisely understand a number being represented on a number line, the first step is to identify how the number line is divided up. When utilizing a number line to represent decimal portions of numbers, it is helpful to label the divisions, or insert additional divisions, as needed.

For example, what number, to the nearest hundredths place, is marked by the point on the following number line?

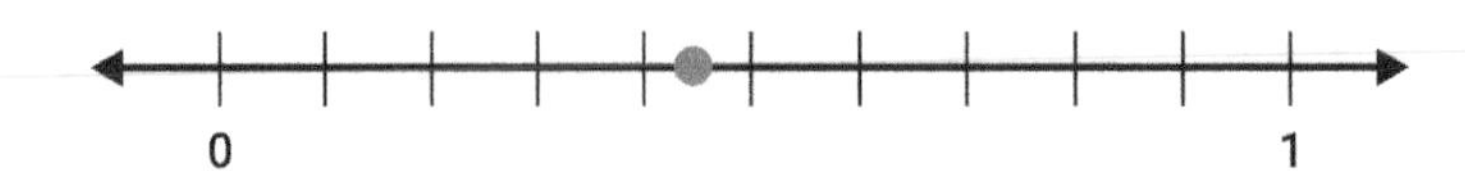

First, figure out how the number line is divided up. In this case, it has ten sections, so it is divided into tenths. To use this number line with the divisions, label the divisions as follows:

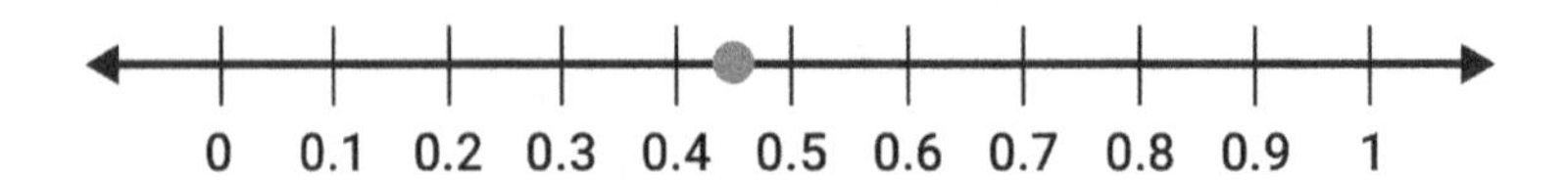

Because the dot is placed equally between 0.4 and 0.5, it is at 0.45.

What number, to the nearest tenths place, is marked by the point on the following number line?

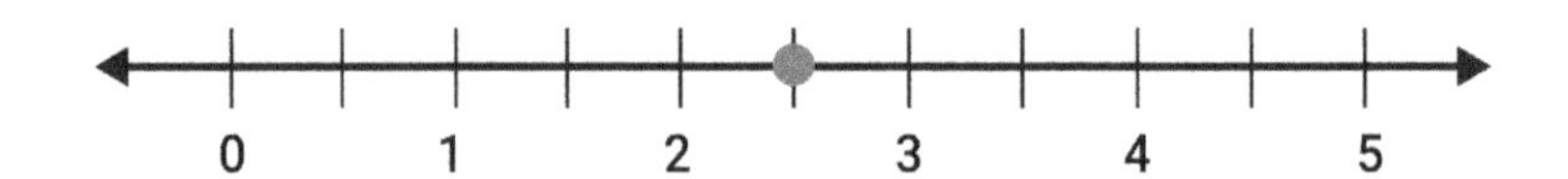

First, determine what the number line is divided up into and mark it on the line.

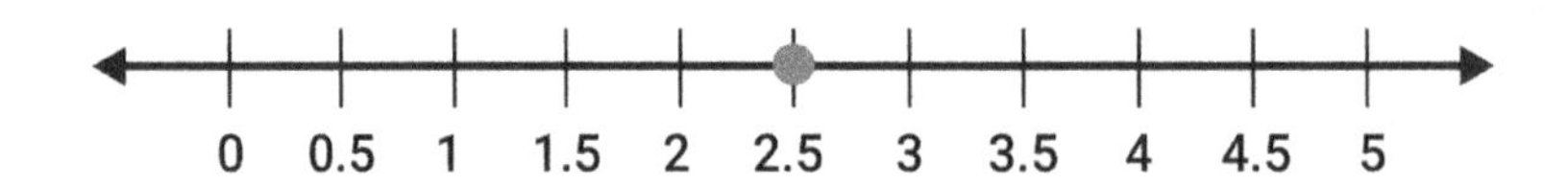

This number line is divided up into half of the whole numbers it represents, or 0.5 increments. The division and labeling of the number line assists in easily reading the dot as marking 2.5.

Ratio Reasoning

Using Ratios to Describe a Relationship Between Two Quantities

Ratios are used to show the relationship between two quantities. The ratio of oranges to apples in the grocery store may be 3 to 2. That means that for every 3 oranges, there are 2 apples. This comparison can be expanded to represent the actual number of oranges and apples, such as 36 oranges to 24 apples. Another example may be the number of boys to girls in a math class. If the ratio of boys to girls is given as 2 to 5, that means there are 2 boys to every 5 girls in the class. Ratios can also be compared if the units in each ratio are the same. The ratio of boys to girls in the math class can be compared to the ratio of boys to girls in a science class by stating which ratio is higher and which is lower.

Rates are used to compare two quantities with different units. **Unit rates** are the simplest form of rate. With unit rates, the denominator in the comparison of two units is one. For example, if someone can type at a rate of 1000 words in 5 minutes, then his or her unit rate for typing is:

$$\frac{1000}{5} = 200 \text{ words in one minute}$$

or

200 words per minute

Any rate can be converted into a unit rate by dividing to make the denominator one. 1000 words in 5 minutes has been converted into the unit rate of 200 words per minute.

Computing Unit Rates

Unit rate word problems will ask to calculate the rate or quantity of something in a different value. For example, a problem might say that a car drove a certain number of miles in a certain number of minutes and then ask how many miles per hour the car was traveling. These questions involve solving proportions. Consider the following examples:

1) Alexandra made $96 during the first 3 hours of her shift as a temporary worker at a law office. She will continue to earn money at this rate until she finishes in 5 more hours. How much does Alexandra make per hour? How much will Alexandra have made at the end of the day?

This problem can be solved in two ways. The first is to set up a proportion, as the rate of pay is constant. The second is to determine her hourly rate, multiply the 5 hours by that rate, and then add the $96.

To set up a proportion, put the money already earned over the hours already worked on one side of an equation. The other side has x over 8 hours (the total hours worked in the day). It looks like this:

$$\frac{96}{3} = \frac{x}{8}$$

Now, cross-multiply to get:

$$768 = 3x$$

To get x, divide by 3, which leaves $x = 256$.

Alternatively, as x is the numerator of one of the proportions, multiplying by its denominator will reduce the solution by one step. Thus, Alexandra will make $256 at the end of the day. To calculate her hourly rate, divide the total by 8, giving $32 per hour.

Alternatively, it is possible to figure out the hourly rate by dividing $96 by 3 hours to get $32 per hour. Now her total pay can be figured by multiplying $32 per hour by 8 hours, which comes out to $256.

2) Jonathan is reading a novel. So far, he has read 215 of the 335 total pages. It takes Jonathan 25 minutes to read 10 pages, and the rate is constant. How long does it take Jonathan to read one page? How much longer will it take him to finish the novel? Express the answer in time.

To calculate how long it takes Jonathan to read one page, divide the 25 minutes by 10 pages to determine the page per minute rate. Thus, it takes 2.5 minutes to read one page.

Jonathan must read 120 more pages to complete the novel. (This is calculated by subtracting the pages already read from the total.) Now, multiply his rate per page by the number of pages. Thus:

$$120 \times 2.5 = 300$$

Expressed in time, 300 minutes is equal to 5 hours.

3) At a hotel, $\frac{4}{5}$ of the 120 rooms are booked for Saturday. On Sunday, $\frac{3}{4}$ of the rooms are booked. On which day are more of the rooms booked, and by how many more?

The first step is to calculate the number of rooms booked for each day. Do this by multiplying the fraction of the rooms booked by the total number of rooms.

Saturday: $\frac{4}{5} \times 120 = \frac{4}{5} \times \frac{120}{1} = \frac{480}{5} = 96$ rooms

Sunday: $\frac{3}{4} \times 120 = \frac{3}{4} \times \frac{120}{1} = \frac{360}{4} = 90$ rooms

Thus, more rooms were booked on Saturday by 6 rooms.

4) In a veterinary hospital, the veterinarian-to-pet ratio is 1:9. The ratio is always constant. If there are 45 pets in the hospital, how many veterinarians are currently in the veterinary hospital?

Set up a proportion to solve for the number of veterinarians:

$$\frac{1}{9} = \frac{x}{45}$$

Cross-multiplying results in $9x = 45$, which works out to 5 veterinarians.

Alternatively, as there are always 9 times as many pets as veterinarians, it is possible to divide the number of pets (45) by 9. This also arrives at the correct answer of 5 veterinarians.

5) At a general practice law firm, 30% of the lawyers work solely on tort cases. If 9 lawyers work solely on tort cases, how many lawyers work at the firm?

First, solve for the total number of lawyers working at the firm, which will be represented here with x. The problem states that 9 lawyers work solely on torts cases, and they make up 30% of the total lawyers at the firm. Thus, 30% multiplied by the total, x, will equal 9. Written as equation, this is:

$$30\% \times x = 9$$

It's easier to deal with the equation after converting the percentage to a decimal, leaving:

$$0.3x = 9$$

Thus:

$$x = \frac{9}{0.3} = 30$$

lawyers working at the firm.

6) Xavier was hospitalized with pneumonia. He was originally given 35mg of antibiotics. Later, after his condition continued to worsen, Xavier's dosage was increased to 60mg. What was the percent increase of the antibiotics? Round the percentage to the nearest tenth.

An increase or decrease in percentage can be calculated by dividing the difference in amounts by the original amount and multiplying by 100. Written as an equation, the formula is:

$$\frac{new\ quantity - old\ quantity}{old\ quantity} \times 100$$

Here, the question states that the dosage was increased from 35mg to 60mg, so these are plugged into the formula to find the percentage increase.

$$\frac{60 - 35}{35} \times 100 = \frac{25}{35} \times 100 = .7142 \times 100 = 71.4\%$$

Using Ratio Reasoning to Convert Rates

Ratios and rates can be used together to convert rates into different units. For example, if someone is driving 50 kilometers per hour, that rate can be converted into miles per hour by using a ratio known as the **conversion factor**. Since the given value contains kilometers and the final answer needs to be in miles, the ratio relating miles to kilometers needs to be used. There are 0.62 miles in 1 kilometer. This, written as a ratio and in fraction form, is:

$$\frac{0.62\ miles}{1\ km}$$

To convert 50km/hour into miles per hour, the following conversion needs to be set up:

$$\frac{50\ km}{hour} \times \frac{0.62\ miles}{1\ km} = 31\ miles\ per\ hour$$

Solving Problems Involving Scale Factors

The ratio between two similar geometric figures is called the **scale factor**. For example, a problem may depict two similar triangles, A and B. The scale factor from the smaller triangle A to the larger triangle B is given as 2 because the length of the corresponding side of the larger triangle, 16, is twice the corresponding side on the smaller triangle, 8. This scale factor can also be used to find the value of a missing side, x, in triangle A. Since the scale factor from the smaller triangle (A) to larger one (B) is 2, the larger corresponding side in triangle B (given as 25), can be divided by 2 to find the missing side in A (x = 12.5). The scale factor can also be represented in the equation $2A = B$ because two times the lengths of A gives the corresponding lengths of B. This is the idea behind similar triangles.

Proportional Relationships

Much like a scale factor can be written using an equation like $2A = B$, a **relationship** is represented by the equation $Y = kX$. X and Y are **proportional** because as values of X increase, the values of Y also increase. A relationship that is **inversely proportional** can be represented by the equation $Y = \frac{k}{X}$, where the value of Y decreases as the value of x increases and vice versa.

Proportional reasoning can be used to solve problems involving ratios, percentages, and averages. Ratios can be used in setting up proportions and solving them to find unknowns. For example, if a student completes an average of 10 pages of math homework in 3 nights, how long would it take the student to complete 22 pages? Both ratios can be written as fractions. The second ratio would contain the unknown.

The following proportion represents this problem, where x is the unknown number of nights:

$$\frac{10\ pages}{3\ nights} = \frac{22\ pages}{x\ nights}$$

Solving this proportion entails cross-multiplying and results in the following equation:

$$10x = 22 \times 3$$

Simplifying and solving for x results in the exact solution:

$$x = 6.6\ nights$$

The result would be rounded up to 7 because the homework would actually be completed on the 7th night.

The following problem uses ratios involving percentages:

If 20% of the class is girls and 30 students are in the class, how many girls are in the class?

To set up this problem, it is helpful to use the common proportion:

$$\frac{\%}{100} = \frac{is}{of}$$

Within the proportion, % is the percentage of girls, 100 is the total percentage of the class, *is* is the number of girls, and *of* is the total number of students in the class. Most percentage problems can be written using this language. To solve this problem, the proportion should be set up as

$$\frac{20}{100} = \frac{x}{30}$$

and then solved for x. Cross-multiplying results in the equation:

$$20 \times 30 = 100x$$

which results in the solution $x = 6$. There are 6 girls in the class.

Problems involving volume, length, and other units can also be solved using ratios. For example, a problem may ask for the volume of a cone to be found that has a radius, $r = 7m$ and a height, $h = 16m$. Referring to the formulas provided on the test, the volume of a cone is given as:

$$V = \pi r^2 \frac{h}{3}$$

where r is the radius, and h is the height. Plugging $r = 7$ and $h = 16$ into the formula, the following is obtained:

$$V = \pi(7^2)\frac{16}{3}$$

Therefore, volume of the cone is found to be approximately $821m^3$. Sometimes, answers in different units are sought. If this problem wanted the answer in liters, $821m^3$ would need to be converted. Using the equivalence statement:

$$1m^3 = 1000L$$

the following ratio would be used to solve for liters:

$$821\text{m}^3 \times \frac{1000L}{1m^3}$$

Cubic meters in the numerator and denominator cancel each other out, and the answer is converted to:

821,000 liters

or

$$8.21 \times 10^5 \text{ L}$$

Other conversions can also be made between different given and final units. If the temperature in a pool is 30°C, what is the temperature of the pool in degrees Fahrenheit? To convert these units, an equation is used relating Celsius to Fahrenheit. The following equation is used:

$$T_{°F} = 1.8T_{°C} + 32$$

Plugging in the given temperature and solving the equation for T yields the result:

$$T_{°F} = 1.8(30) + 32 = 86°F$$

Both units in the metric system and U.S. customary system are widely used.

Here are some more examples of how to solve for proportions:

1) $\frac{75\%}{90\%} = \frac{25\%}{x}$

To solve for x, the fractions must be cross multiplied:

$$75\%x = 90\% \times 25\%$$

To make things easier, let's convert the percentages to decimals:

$$0.9 \times 0.25 = 0.225 = 0.75x$$

To get rid of x's coefficient, each side must be divided by that same coefficient to get the answer:

$$x = 0.3$$

The question could ask for the answer as a percentage or fraction in lowest terms, which are 30% and $\frac{3}{10}$, respectively.

2) $\frac{x}{12} = \frac{30}{96}$

Cross-multiply: $96x = 30 \times 12$

Multiply: $96x = 360$

Divide: $x = 360 \div 96$

Answer: $x = 3.75$

3) $\frac{0.5}{3} = \frac{x}{6}$

Cross-multiply: $3x = 0.5 \times 6$

Multiply: $3x = 3$

Divide: $x = 3 \div 3$

Answer: $x = 1$

You may have noticed there's a faster way to arrive at the answer. If there is an obvious operation being performed on the proportion, the same operation can be used on the other side of the proportion to solve for x. For example, in the first practice problem, 75% became 25% when divided by 3, and upon doing the same to 90%, the correct answer of 30% would have been found with much less legwork. However, these questions aren't always so intuitive, so it's a good idea to work through the steps, even if the answer seems apparent from the outset.

Solving Ratio and Percent Problems

Questions dealing with percentages can be difficult when they are phrased as word problems. These word problems almost always come in three varieties. The first type will ask to find what percentage of some number will equal another number. The second asks to determine what number is some percentage of another given number. The third will ask what number another number is a given percentage of.

One of the most important parts of correctly answering percentage word problems is to identify the numerator and the denominator. This fraction can then be converted into a percentage, as described above.

The following word problem shows how to make this conversion:

> A department store carries several different types of footwear. The store is currently selling 8 athletic shoes, 7 dress shoes, and 5 sandals. What percentage of the store's footwear are sandals?

First, calculate what serves as the "whole," as this will be the denominator. How many total pieces of footwear does the store sell? The store sells 20 different types (8 athletic + 7 dress + 5 sandals).

Second, what footwear type is the question specifically asking about? Sandals. Thus, 5 is the numerator.

Third, the resultant fraction must be expressed as a percentage. The first two steps indicate that $\frac{5}{20}$ of the footwear pieces are sandals. This fraction must now be converted into a percentage:

$$\frac{5}{20} \times \frac{5}{5} = \frac{25}{100} = 25\%$$

Exponents

Exponents are used in mathematics to express a number or variable multiplied by itself a certain number of times. For example, x^3 means x is multiplied by itself three times. In this expression, x is called the **base**, and 3 is the **exponent.** Exponents can be used in more complex problems when they contain fractions and negative numbers.

Fractional exponents can be explained by looking first at the inverse of exponents, which are **roots.** Given the expression x^2, the square root can be taken, $\sqrt{x^2}$, cancelling out the 2 and leaving *x* by itself, if *x* is positive. Cancellation occurs because $\sqrt{x}$ can be written with exponents, instead of roots, as $x^{\frac{1}{2}}$. The numerator of 1 is the exponent, and the denominator of 2 is called the **root** (which is why it's referred to as a **square root**). Taking the square root of x^2 is the same as raising it to the $\frac{1}{2}$ power. Written out in mathematical form, it takes the following progression:

$$\sqrt{x^2} = (x^2)^{\frac{1}{2}} = x$$

From properties of exponents:

$$2 \times \frac{1}{2} = 1$$

is the actual exponent of *x*. Another example can be seen with $x^{\frac{4}{7}}$. The variable *x*, raised to four-sevenths, is equal to the seventh root of *x* to the fourth power: $\sqrt[7]{x^4}$. In general,

$$x^{\frac{1}{n}} = \sqrt[n]{x}$$

and

$$x^{\frac{m}{n}} = \sqrt[n]{x^m}$$

Negative exponents also involve fractions. Whereas y^3 can also be rewritten as $\frac{y^3}{1}$, y^{-3} can be rewritten as $\frac{1}{y^3}$. A negative exponent means the exponential expression must be moved to the opposite spot in a fraction to make the exponent positive. If the negative appears in the numerator, it moves to the denominator. If the negative appears in the denominator, it is moved to the numerator. In general, $a^{-n} = \frac{1}{a^n}$, and a^{-n} and a^n are reciprocals.

Take, for example, the following expression:

$$\frac{a^{-4}b^2}{c^{-5}}$$

Since *a* is raised to the negative fourth power, it can be moved to the denominator. Since *c* is raised to the negative fifth power, it can be moved to the numerator. The *b* variable is raised to the positive second power, so it does not move.

The simplified expression is as follows:

$$\frac{b^2c^5}{a^4}$$

In mathematical expressions containing exponents and other operations, the order of operations must be followed. **PEMDAS** states that exponents are calculated after any parenthesis and grouping symbols but before any multiplication, division, addition, and subtraction.

Scientific Notation

Scientific notation is used to represent numbers that are either very small or very large. For example, the distance to the Sun is approximately 150,000,000,000 meters. Instead of writing this number with so many zeros, it can be written in scientific notation as 1.5×10^{11}meters. The same is true for very small numbers, but the exponent becomes negative. If the mass of a human cell is 0.000000000001 kilograms, that measurement can be easily represented by 1.0×10^{-12} kilograms. In both situations, scientific notation makes the measurement easier to read and understand. Each number is translated to an expression with one digit in the tens place multiplied by an expression corresponding to the zeros.

When two measurements are given and both involve scientific notation, it is important to know how these interact with each other:

- In addition and subtraction, the exponent on the ten must be the same before any operations are performed on the numbers. For example, $(1.3 \times 10^4) + (3.0 \times 10^3)$ cannot be added until one of the exponents on the ten is changed. The 3.0×10^3 can be changed to 0.3×10^4, then the 1.3 and 0.3 can be added. The answer comes out to be 1.6×10^4.

- For multiplication, the first numbers can be multiplied and then the exponents on the tens can be added. Once an answer is formed, it may have to be converted into scientific notation again depending on the change that occurred.

- The following is an example of multiplication with scientific notation:

$$(4.5 \times 10^3) \times (3.0 \times 10^{-5}) = 13.5 \times 10^{-2}$$

- Since this answer is not in scientific notation, the decimal is moved over to the left one unit, and 1 is added to the ten's exponent. This results in the final answer:

$$1.35 \times 10^{-1}$$

- For division, the first numbers are divided, and the exponents on the tens are subtracted. Again, the answer may need to be converted into scientific notation form, depending on the type of changes that occurred during the problem.
- **Order of magnitude** relates to scientific notation and is the total count of powers of 10 in a number. For example, there are 6 orders of magnitude in 1,000,000. If a number is raised by an order of magnitude, it is multiplied by 10. Order of magnitude can be helpful in estimating results using very large or small numbers. An answer should make sense in terms of its order of magnitude.
- For example, if area is calculated using two dimensions with 6 orders of magnitude, because area involves multiplication, the answer should have around 12 orders of magnitude. Also, answers can be estimated by rounding to the largest place value in each number. For example:

$$5{,}493{,}302 \times 2{,}523{,}100$$

can be estimated by $5 \times 3 = 15$ with 6 orders of magnitude.

Simplifying and Approximating Radicals

The **square root symbol** is expressed as $\sqrt{\ }$ and is commonly known as the **radical.** Taking the root of a number is the inverse operation of multiplying that number by itself some amount of times. For example, squaring the number 7 is equal to 7 × 7, or 49. Finding the square root is the opposite of finding an exponent, as the operation seeks a number that when multiplied by itself equals the number in the square root symbol.

For example:

$$\sqrt{36} = 6$$

because 6 multiplied by 6 equals 36. Note, the square root of 36 is also -6 since:

$$-6 \times -6 = 36$$

This can be indicated using a plus/minus symbol like this: ± 6. However, square roots are often just expressed as a positive number for simplicity with it being understood that the true value can be either positive or negative.

Perfect squares are numbers with whole number square roots. The list of perfect squares begins with 0, 1, 4, 9, 16, 25, 36, 49, 64, 81, and 100.

Determining the square root of imperfect squares requires a calculator to reach an exact figure. It's possible, however, to approximate the answer by finding the two perfect squares that the number fits between. For example, the square root of 40 is between 6 and 7 since the squares of those numbers are 36 and 49, respectively.

Square roots are the most common root operation. If the radical doesn't have a number to the upper left of the symbol, $\sqrt{\ }$, then it's a square root. Sometimes a radical includes a number in the upper left, like $\sqrt[3]{27}$, as in the other common root type—the cube root. Complicated roots like the cube root often require a calculator.

Determining the Reasonableness of Results

The Reasonableness of Results

When solving math word problems, the solution obtained should make sense within the given scenario. The step of checking the solution will reduce the possibility of a calculation error or a solution that may be mathematically correct but not applicable in the real world. Consider the following scenarios:

A problem states that Lisa got 24 out of 32 questions correct on a test and asks to find the percentage of correct answers. To solve the problem, a student divided 32 by 24 to get 1.33, and then multiplied by 100 to get 133 percent. By examining the solution within the context of the problem, the student should recognize that getting all 32 questions correct will produce a perfect score of 100 percent. Therefore, a score of 133 percent with 8 incorrect answers does not make sense, and the calculations should be checked.

A problem states that the maximum weight on a bridge cannot exceed 22,000 pounds. The problem asks to find the maximum number of cars that can be on the bridge at one time if each car weighs 4,000 pounds. To solve this problem, a student divided 22,000 by 4,000 to get an answer of 5.5. By examining the solution within the context of the problem, the student should recognize that although the calculations are mathematically correct, the solution does not make sense. Half of a car on a bridge is not possible, so the student should determine that a maximum of 5 cars can be on the bridge at the same time.

Estimating

Estimation is finding a value that is close to a solution but is not the exact answer. For example, if there are values in the thousands to be multiplied, then each value can be estimated to the nearest thousand and the calculation performed. This value provides an approximate solution that can be determined very quickly.

Rounding is the process of either bumping a number up or leaving it the same, based on a specified place value. First, the place value is specified. Then, the digit to its right is looked at. For example, if rounding to the nearest hundreds place, the digit in the tens place is used. If it is a 0, 1, 2, 3, or 4, the digit being rounded to is left alone. If it is a 5, 6, 7, 8 or 9, the digit being rounded to is increased by one. All other digits before the decimal point are then changed to zeros, and the digits in decimal places are dropped. If a decimal place is being rounded to, all subsequent digits are just dropped. For example, if 845,231.45 was to be rounded to the nearest thousands place, the answer would be 845,000. The 5 would remain the same due to the 2 in the hundreds place. Also, if 4.567 was to be rounded to the nearest tenths place, the answer would be 4.6. The 5 increased to 6 due to the 6 in the hundredths place, and the rest of the decimal is dropped.

Sometimes when performing operations such as multiplying numbers, the result can be estimated by rounding. For example, to estimate the value of:

$$11.2 \times 2.01$$

each number can be rounded to the nearest integer. This will yield a result of 22.

Rounding numbers helps with estimation because it changes the given number to a simpler, although less accurate, number than the exact given number. Rounding allows for easier calculations, which estimate the results of using the exact given number. The accuracy of the estimate and ease of use depends on the place value to which the number is rounded. Rounding numbers consists of:

- determining what place value the number is being rounded to
- examining the digit to the right of the desired place value to decide whether to round up or keep the digit, and
- replacing all digits to the right of the desired place value with zeros.

To round 746,311 to the nearest ten thousand, the digit in the ten thousands place should be located first. In this case, this digit is 4 (7<u>4</u>6,311). Then, the digit to its right is examined. If this digit is 5 or greater, the number will be rounded up by increasing the digit in the desired place by one. If the digit to the right of the place value being rounded is 4 or less, the number will be kept the same. For the given example, the digit being examined is a 6, which means that the number will be rounded up by increasing the digit to the left by one. Therefore, the digit 4 is changed to a 5. Finally, to write the rounded number, any digits to the left of the place value being rounded remain the same and any to its right are replaced with zeros. For the given example, rounding 746,311 to the nearest ten thousand will produce 750,000. To round 746,311 to the nearest hundred, the digit to the right of the three in the hundreds place is examined to determine whether to round up or keep the same number. In this case, that digit is a 1, so the number will be kept the same and any digits to its right will be replaced with zeros. The resulting rounded number is 746,300.

Rounding place values to the right of the decimal follows the same procedure, but digits being replaced by zeros can simply be dropped. To round 3.752891 to the nearest thousandth, the desired place value is located (3.75<u>2</u>891), and the digit to the right is examined. In this case, the digit 8 indicates that the number will be rounded up, and the 2 in the thousandths place will increase to a 3. Rounding up and replacing the digits to the right of the thousandths place produces 3.753000 which is equivalent to 3.753. Therefore, the zeros are not necessary, and the rounded number should be written as 3.753.

When rounding up, if the digit to be increased is a 9, the digit to its left is increased by 1 and the digit in the desired place value is changed to a zero. For example, the number 1,598 rounded to the nearest ten is 1,600. Another example shows the number 43.72961 rounded to the nearest thousandth is 43.730 or 43.73.

Mental math should always be considered as problems are worked through, and the ability to work through problems in one's head helps save time. If a problem is simple enough, such as $15 + 3 = 18$, it should be completed mentally. The ability to do this will increase once addition and subtraction in higher place values are grasped. Also, mental math is important in multiplication and division. The times tables multiplying all numbers from 1 to 12 should be memorized. This will allow for division within those numbers to be memorized as well. For example, we should know easily that:

$$121 \div 11 = 11$$

because it should be memorized that $11 \times 11 = 121$. Here is the multiplication table to be memorized:

x	1	2	3	4	5	6	7	8	9	10	11	12	13	14	15
1	1	2	3	4	5	6	7	8	9	10	11	12	13	14	15
2	2	4	6	8	10	12	14	16	18	20	22	24	26	28	30
3	3	6	9	12	15	18	21	24	27	30	33	36	39	42	45
4	4	8	12	16	20	24	28	32	36	40	44	48	52	56	60
5	5	10	15	20	25	30	35	40	45	50	55	60	65	70	75
6	6	12	18	24	30	36	42	48	54	60	66	72	78	84	90
7	7	14	21	28	35	42	49	56	63	70	77	84	91	98	105
8	8	16	24	32	40	48	56	64	72	80	88	96	104	112	120
9	9	18	27	36	45	54	63	72	81	90	99	108	117	126	135
10	10	20	30	40	50	60	70	80	90	100	110	120	130	140	150
11	11	22	33	44	55	66	77	88	99	110	121	132	143	154	165
12	12	24	36	48	60	72	84	96	108	120	132	144	156	168	180
13	13	26	39	52	65	78	91	104	117	130	143	156	169	182	195
14	14	28	42	56	70	84	98	112	126	140	154	168	182	196	210
15	15	30	45	60	75	90	105	120	135	150	165	180	195	210	225

The values along the diagonal of the table consist of **perfect squares**. A perfect square is a number that represents a product of two equal integers.

Estimating Absolute and Relative Error in the Numerical Answer to a Problem

Once a result is determined to be logical within the context of a given problem, the result should be evaluated by its nearness to the expected answer. This is performed by approximating given values to perform mental math. Numbers should be rounded to the nearest value possible to check the initial results.

Consider the following example:

> A customer is buying a new sound system for their home. The customer purchases a stereo for \$435, 2 speakers for \$67 each, and the necessary cables for \$12. The customer chooses an option that allows him to spread the costs over equal payments for 4 months. How much will the monthly payments be?

After making calculations for the problem, a student determines that the monthly payment will be \$145.25. To check the accuracy of the results, the student rounds each cost to the nearest ten:

$$(440 + 70 + 70 + 10)$$

and determines that the total is approximately \$590. Dividing by 4 months gives an approximate monthly payment of \$147.50. Therefore, the student can conclude that the solution of \$145.25 is very close to what should be expected.

When rounding, the place-value that is used in rounding can make a difference. Suppose the student had rounded to the nearest hundred for the estimation. The result:

$$400 + 100 + 100 + 0 = 600$$

$$600 \div 4 = 150$$

will show that the answer is reasonable but not as close to the actual value as rounding to the nearest ten.

When considering the accuracy of estimates, the error in the estimated solution can be shown as absolute and relative. **Absolute error** tells the actual difference between the estimated value and the true, calculated value. The **relative error** tells how large the error is in relation to the true value. There may be two problems where the absolute error of the values (the estimated one and the calculated one) is 10. For one problem, this may mean the relative error in the estimate is very small because the estimated value is 15,000, and the true value is 14,990. Ten in relation to the true value of 15,000 is small: 0.06%. For the other problem, the estimated value is 50, and the true value is 40. In this case, the absolute error of 10 means a high relative error because the true value is smaller. The relative error is:

$$\frac{10}{40} = 0.25 \text{ or } 25\%$$

Algebra

Algebraic Expressions and Equations

An **algebraic expression** is a statement about an unknown quantity expressed in mathematical symbols. A variable is used to represent the unknown quantity, usually denoted by a letter. An equation is a statement in which two expressions (at least one containing a variable) are equal to each other. An algebraic expression can be thought of as a mathematical phrase and an equation can be thought of as a mathematical sentence.

Algebraic expressions and equations both contain numbers, variables, and mathematical operations. The following are examples of algebraic expressions:

$$5x + 3, 7xy - 8(x^2 + y)$$

and

$$\sqrt{a^2 + b^2}$$

An expression can be simplified or evaluated for given values of variables. The following are examples of equations:

$$2x + 3 = 7$$

$$a^2 + b^2 = c^2$$

$$2x + 5 = 3x - 2$$

An equation contains two sides separated by an equal sign. **Equations** can be solved to determine the value(s) of the variable for which the statement is true.

Adding and Subtracting Linear Algebraic Expressions

An algebraic expression is simplified by combining like terms. A **term** is a number, variable, or product of a number and variables separated by addition and subtraction. For the algebraic expression:

$$3x^2 - 4x + 5 - 5x^2 + x - 3$$

the terms are $3x^2$, -4x, 5, -$5x^2$, x, and -3. Like terms have the same variables raised to the same powers (exponents). The like terms for the previous example are $3x^2$ and -$5x^2$, -4x and x, 5 and -3. To combine like terms, the coefficients (numerical factor of the term including sign) are added, and the variables and their powers are kept the same. Note that if a coefficient is not written, it is an implied coefficient of 1 ($x = 1x$). The previous example will simplify to:

$$-2x^2 - 3x + 2$$

When adding or subtracting algebraic expressions, each expression is written in parenthesis. The negative sign is distributed when necessary, and like terms are combined. Consider the following: add $2a + 5b - 2$ to $a - 2b + 8c - 4$. The sum is set as follows:

$$(a - 2b + 8c - 4) + (2a + 5b - 2)$$

In front of each set of parentheses is an implied positive one, which, when distributed, does not change any of the terms. Therefore, the parentheses are dropped and like terms are combined:

$$a - 2b + 8c - 4 + 2a + 5b - 2$$

$$3a + 3b + 8c - 6$$

Consider the following problem: Subtract $2a + 5b - 2$ from $a - 2b + 8c - 4$. The difference is set as follows:

$$(a - 2b + 8c - 4) - (2a + 5b - 2)$$

The implied one in front of the first set of parentheses will not change those four terms. However, distributing the implied -1 in front of the second set of parentheses will change the sign of each of those three terms:

$$a - 2b + 8c - 4 - 2a - 5b + 2$$

Combining like terms yields the simplified expression:

$$-a - 7b + 8c - 2$$

Distributive Property

The **distributive property** states that multiplying a sum (or difference) by a number produces the same result as multiplying each value in the sum (or difference) by the number and adding (or subtracting) the products. Using mathematical symbols, the distributive property states:

$$a(b + c) = ab + ac$$

The expression:

$$4(3+2)$$

is simplified using the order of operations. Simplifying inside the parenthesis first produces 4×5, which equals 20. The expression:

$$4(3+2)$$

can also be simplified using the distributive property:

$$4(3+2) = 4 \times 3 + 4 \times 2 = 12 + 8 = 20$$

Consider the following example:

$$4(3x-2)$$

The expression cannot be simplified inside the parenthesis because $3x$ and -2 are not like terms and therefore cannot be combined. However, the expression can be simplified by using the distributive property and multiplying each term inside of the parenthesis by the term outside of the parenthesis:

$$12x-8$$

The resulting equivalent expression contains no like terms, so it cannot be further simplified.

Consider the expression:

$$(3x+2y+1)-(5x-3)+2(3y+4)$$

Again, there are no like terms, but the distributive property is used to simplify the expression. Note there is an implied one in front of the first set of parentheses and an implied -1 in front of the second set of parentheses. Distributing the 1, -1, and 2 produces:

$$1(3x)+1(2y)+1(1)-1(5x)-1(-3)+2(3y)+2(4)$$

$$3x+2y+1-5x+3+6y+8$$

This expression contains like terms that are combined to produce the simplified expression:

$$-2x+8y+12$$

Algebraic expressions are tested to be equivalent by choosing values for the variables and evaluating both expressions. For example:

$$4(3x-2)$$

and $12x-8$ are tested by substituting 3 for the variable x and calculating to determine if equivalent values result.

Simple Expressions for Given Values

An **algebraic expression** is a statement written in mathematical symbols, typically including one or more unknown values represented by variables. For example, the expression:

$$2x + 3$$

states that an unknown number (*x*) is multiplied by 2 and added to 3. If given a value for the unknown number, or variable, the value of the expression is determined. For example, if the value of the variable *x* is 4, the value of the expression 4 is multiplied by 2, and 3 is added. This results in a value of 11 for the expression.

When given an algebraic expression and values for the variable(s), the expression is evaluated to determine its numerical value. To evaluate the expression, the given values for the variables are substituted (or replaced), and the expression is simplified using the order of operations. Parenthesis should be used when substituting. Consider the following: Evaluate:

$$a - 2b + ab$$

for

$$a = 3$$

and

$$b = -1$$

To evaluate, any variable *a* is replaced with 3 and any variable *b* with -1, producing:

$$(3) - 2(-1) + (3)(-1)$$

Next, the order of operations is used to calculate the value of the expression, which is 2.

Parts of Expressions

Algebraic expressions consist of variables, numbers, and operations. A term of an expression is any combination of numbers and/or variables, and terms are separated by addition and subtraction. For example, the expression:

$$5x^2 - 3xy + 4 - 2$$

consists of 4 terms: $5x^2$, -3*xy*, 4*y*, and -2. Note that each term includes its given sign (+ or —). The variable part of a term is a letter that represents an unknown quantity. The coefficient of a term is the number by which the variable is multiplied. For the term 4*y*, the variable is y, and the coefficient is 4. Terms are identified by the power (or exponent) of its variable.

A number without a variable is referred to as a **constant**. If the variable is to the first power (x^1 or simply *x*), it is referred to as a **linear term**. A term with a variable to the second power (x^2) is quadratic, and a term to the third power (x^3) is cubic. Consider the expression:

$$x^3 + 3x - 1$$

The constant is -1. The linear term is 3*x*. There is no quadratic term. The cubic term is x^3.

An algebraic expression can also be classified by how many terms exist in the expression. Any like terms should be combined before classifying. A monomial is an expression consisting of only one term. Examples of monomials are: 17, 2*x*, and $-5ab^2$. A binomial is an expression consisting of two terms separated by addition or subtraction. Examples include:

$$2x - 4$$

and

$$-3y^2 + 2y$$

A trinomial consists of 3 terms. For example, $5x^2 - 2x + 1$ is a trinomial.

Verbal Statements and Algebraic Expressions

As mentioned, an algebraic expression is a statement about unknown quantities expressed in mathematical symbols. The statement *five times a number added to forty* is expressed as:

$$5x + 40$$

An equation is a statement in which two expressions (with at least one containing a variable) are equal to one another. The statement *five times a number added to forty is equal to ten* is expressed as:

$$5x + 40 = 10$$

Real world scenarios can also be expressed mathematically. Suppose a job pays its employees $300 per week and $40 for each sale made. The weekly pay is represented by the expression $40x + 300$ where *x* is the number of sales made during the week.

Consider the following scenario:

> Bob had $20 and Tom had $4. After selling 4 ice cream cones to Bob, Tom has as much money as Bob. The cost of an ice cream cone is an unknown quantity and can be represented by a variable (*x*). The amount of money Bob has after his purchase is four times the cost of an ice cream cone subtracted from his original $\$20 \rightarrow 20 - 4x$. The amount of money Tom has after his sale is four times the cost of an ice cream cone added to his original $\$4 \rightarrow 4x + 4$. After the sale, the amount of money that Bob and Tom have is equal $\rightarrow 20 - 4x = 4x + 4$.

When expressing a verbal or written statement mathematically, it is vital to understand words or phrases that can be represented with symbols. The following are examples:

Symbol	Phrase
+	Added to; increased by; sum of; more than
–	Decreased by; difference between; less than; take away
×	Multiplied by; 3(4,5...) times as large; product of
÷	Divided by; quotient of; half (third, etc.) of
=	Is; the same as; results in; as much as; equal to
x,t,n, etc.	A number; unknown quantity; value of; variable

Use of Formulas

Formulas are mathematical expressions that define the value of one quantity, given the value of one or more different quantities. Formulas look like equations because they contain variables, numbers, operators, and an equal sign. All formulas are equations, but not all equations are formulas. A formula must have more than one variable. For example:

$$2x + 7 = y$$

is an equation and a formula (it relates the unknown quantities x and y). However:

$$2x + 7 = 3$$

is an equation but not a formula (it only expresses the value of the unknown quantity x).

Formulas are typically written with one variable alone (or isolated) on one side of the equal sign. This variable can be thought of as the *subject* in that the formula is stating the value of the *subject* in terms of the relationship between the other variables. Consider the distance formula:

$$distance = rate \times time$$

or

$$d = rt$$

The value of the subject variable d (distance) is the product of the variable r and t (rate and time). Given the rate and time, the distance traveled can easily be determined by substituting the values into the formula and evaluating.

The formula:

$$P = 2l + 2w$$

expresses how to calculate the perimeter of a rectangle (P) given its length (l) and width (w). To find the perimeter of a rectangle with a length of 3ft and a width of 2ft, these values are substituted into the formula for l and w:

$$P = 2(3ft) + 2(2ft)$$

Following the order of operations, the perimeter is determined to be 10ft. When working with formulas such as these, including units is an important step.

Given a formula expressed in terms of one variable, the formula can be manipulated to express the relationship in terms of any other variable. In other words, the formula can be rearranged to change which variable is the *subject*. To solve for a variable of interest by manipulating a formula, the equation may be solved as if all other variables were numbers. The same steps for solving are followed, leaving operations in terms of the variables instead of calculating numerical values. For the formula:

$$P = 2l + 2w$$

the perimeter is the subject expressed in terms of the length and width. To write a formula to calculate the width of a rectangle, given its length and perimeter, the previous formula relating the three variables is solved for the variable *w*. If *P* and *l* were numerical values, this is a two-step linear equation solved by subtraction and division. To solve the equation $P = 2l + 2w$ for *w*, 2*l* is first subtracted from both sides:

$$P - 2l = 2w$$

Then both sides are divided by 2:

$$\frac{P - 2l}{2} = w$$

Dependent and Independent Variables

A variable represents an unknown quantity, and, in the case of a formula, a specific relationship exists between the variables. Within a given scenario, variables are the quantities that are changing. If two variables exist, one is dependent, and one is independent. The value of one variable depends on the other variable. If a scenario describes distance traveled and time traveled at a given speed, distance is **dependent,** and time is **independent**. The distance traveled depends on the time spent traveling. If a scenario describes the cost of a cab ride and the distance traveled, the cost is dependent, and the distance is independent. The cost of a cab ride depends on the distance travelled. Formulas often contain more than two variables and are typically written with the dependent variable alone on one side of the equation. This lone variable is the subject of the statement. If a formula contains three or more variables, one variable is dependent, and the rest are independent. The values of all independent variables are needed to determine the value of the dependent variable.

The formula:

$$P = 2l + 2w$$

expresses the dependent variable *P* in terms of the independent variables, *l* and *w*. The perimeter of a rectangle depends on its length and width. The formula $d = rt$ $(distance = rate \times time)$ expresses the dependent variable *d* in terms of the independent variables, *r* and *t*. The distance traveled depends on the rate (or speed) and the time traveled.

Multistep One-Variable Linear Equations and Inequalities

Linear equations and linear inequalities are both comparisons of two algebraic expressions. However, unlike equations in which the expressions are equal, linear inequalities compare expressions that may be unequal. **Linear equations** typically have one value for the variable that makes the statement true. **Linear inequalities** generally have an infinite number of values that make the statement true.

When solving a linear equation, the desired result requires determining a numerical value for the unknown variable. If given a linear equation involving addition, subtraction, multiplication, or division, working backwards isolates the variable. Addition and subtraction are inverse operations, as are multiplication and division. Therefore, they can be used to cancel each other out.

The first steps to solving linear equations are distributing, if necessary, and combining any like terms on the same side of the equation. Sides of an equation are separated by an **equal** sign. Next, the equation is manipulated to show the variable on one side. Whatever is done to one side of the equation must be done to the other side of the equation to remain equal. Inverse operations are then used to isolate the variable and undo the order of operations backwards. Addition and subtraction are undone, then multiplication and division are undone.

For example, solve $4(t-2)+2t-4=2(9-2t)$

Distributing: $4t-8+2t-4=18-4t$

Combining like terms: $6t-12=18-4t$

Adding $4t$ to each side to move the variable: $10t-12=18$

Adding 12 to each side to isolate the variable: $10t=30$

Dividing each side by 10 to isolate the variable: $t=3$

The answer can be checked by substituting the value for the variable into the original equation, ensuring that both sides calculate to be equal.

Linear inequalities express the relationship between unequal values. More specifically, they describe in what way the values are unequal. A value can be greater than (>), less than (<), greater than or equal to (≥), or less than or equal to (≤) another value.

$$5x+40>65$$

is read as *five times a number added to forty is greater than sixty-five.*

When solving a linear inequality, the solution is the set of all numbers that makes the statement true. The inequality:

$$x+2\geq 6$$

has a solution set of 4 and every number greater than 4 (4.01; 5; 12; 107; etc.). Adding 2 to 4 or any number greater than 4 results in a value that is greater than or equal to 6. Therefore, $x\geq 4$ is the solution set.

To algebraically solve a linear inequality, follow the same steps as those for solving a linear equation. The inequality symbol stays the same for all operations *except* when multiplying or dividing by a negative number. If multiplying or dividing by a negative number while solving an inequality, the relationship reverses (the sign flips). In other words, > switches to < and vice versa. Multiplying or dividing by a positive number does not change the relationship, so the sign stays the same. An example is shown below.

Solve $-2x-8\leq 22$

Add 8 to both sides: $-2x \leq 30$

Divide both sides by -2: $x \geq -15$

Solutions of a linear equation or a linear inequality are the values of the variable that make a statement true. In the case of a linear equation, the solution set (list of all possible solutions) typically consists of a single numerical value. To find the solution, the equation is solved by isolating the variable. For example, solving the equation:

$$3x - 7 = -13$$

produces the solution $x = -2$. The only value for *x* which produces a true statement is -2. This can be checked by substituting -2 into the original equation to check that both sides are equal. In this case:

$$3(-2) - 7 = -13 \rightarrow -13 = -13$$

therefore, -2 is a solution.

Although linear equations generally have one solution, this is not always the case. If there is no value for the variable that makes the statement true, there is no solution to the equation. Consider the equation:

$$x + 3 = x - 1$$

There is no value for *x* in which adding 3 to the value produces the same result as subtracting one from the value. Conversely, if any value for the variable makes a true statement, the equation has an infinite number of solutions. Consider the equation:

$$3x + 6 = 3(x + 2)$$

Any number substituted for *x* will result in a true statement (both sides of the equation are equal).

By manipulating equations like the two above, the variable of the equation will cancel out completely. If the remaining constants express a true statement (ex. $6 = 6$), then all real numbers are solutions to the equation. If the constants left express a false statement (ex. $3 = -1$), then no solution exists for the equation.

Solving a linear inequality requires all values that make the statement true to be determined. For example, solving:

$$3x - 7 \geq -13$$

produces the solution $x \geq -2$. This means that -2 and any number greater than -2 produces a true statement. Solution sets for linear inequalities will often be displayed using a number line. If a value is included in the set (≥ or ≤), a shaded dot is placed on that value and an arrow extending in the direction of the solutions. For a variable > or ≥ a number, the arrow will point right on a number line, the direction where the numbers increase. If a variable is < or ≤ a number, the arrow will point left on a number line, which is the direction where the numbers decrease. If the value is not included in the set (> or <), an open (unshaded) circle on that value is used with an arrow in the appropriate direction.

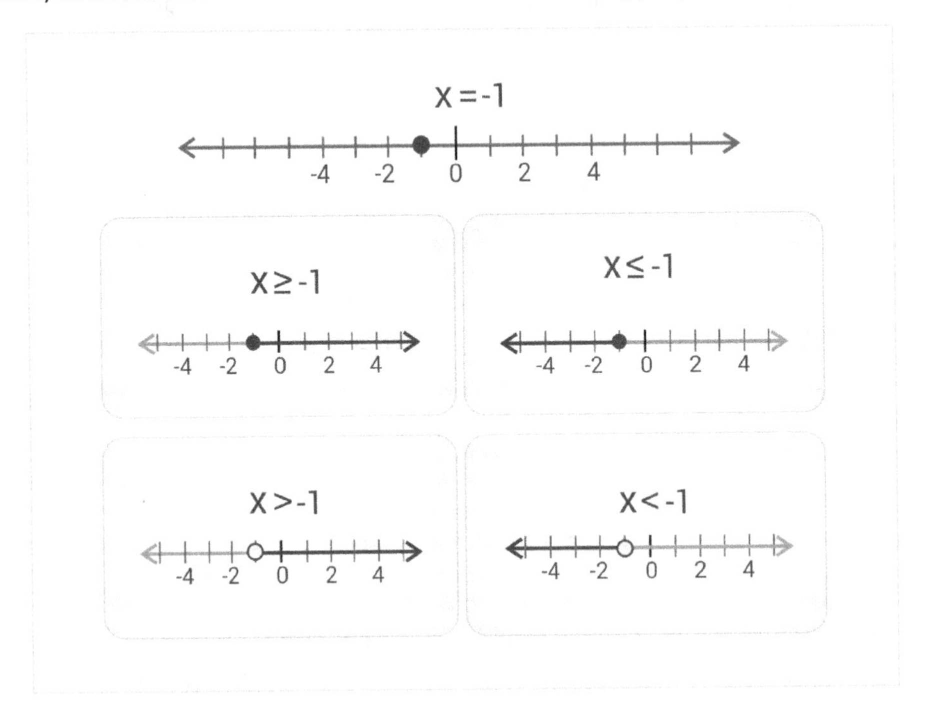

Similar to linear equations, a linear inequality may have a solution set consisting of all real numbers or no solution. When solved algebraically, a linear inequality in which the variable cancels out and results in a true statement (ex. $7 \geq 2$) has a solution set of all real numbers. A linear inequality in which the variable cancels out and results in a false statement (ex. $7 \leq 2$) has no solution.

Linear Relationships

Linear relationships describe the way two quantities change with respect to each other. The relationship is defined as linear because a line is produced if all the sets of corresponding values are graphed on a coordinate grid. When expressing the linear relationship as an equation, the equation is often written in the form:

$$y = mx + b \text{ (slope-intercept form)}$$

where *m* and *b* are numerical values and *x* and *y* are variables (for example, $y = 5x + 10$). Given a linear equation and the value of either variable (*x* or *y*), the value of the other variable can be determined.

Suppose a teacher is grading a test containing 20 questions with 5 points given for each correct answer, adding a curve of 10 points to each test. This linear relationship can be expressed as the equation:

$$y = 5x + 10$$

where *x* represents the number of correct answers, and y represents the test score. To determine the score of a test with a given number of correct answers, the number of correct answers is substituted into the equation for *x* and evaluated. For example, for 10 correct answers, 10 is substituted for *x*:

$$y = 5(10) + 10 \rightarrow y = 60$$

Therefore, 10 correct answers will result in a score of 60. The number of correct answers needed to obtain a certain score can also be determined. To determine the number of correct answers needed to score a 90, 90 is substituted for *y* in the equation (*y* represents the test score) and solved:

$$90 = 5x + 10$$

$$80 = 5x \rightarrow 16 = x$$

Therefore, 16 correct answers are needed to score a 90.

Linear relationships may be represented by a table of 2 corresponding values. Certain tables may determine the relationship between the values and predict other corresponding sets. Consider the table below, which displays the money in a checking account that charges a monthly fee:

Month	0	1	2	3	4
Balance	$210	$195	$180	$165	$150

An examination of the values reveals that the account loses $15 every month (the month increases by one and the balance decreases by 15). This information can be used to predict future values. To determine what the value will be in month 6, the pattern can be continued, and it can be concluded that the balance will be $120. To determine which month the balance will be $0, $210 is divided by $15 (since the balance decreases $15 every month), resulting in month 14.

Similar to a table, a graph can display corresponding values of a linear relationship.

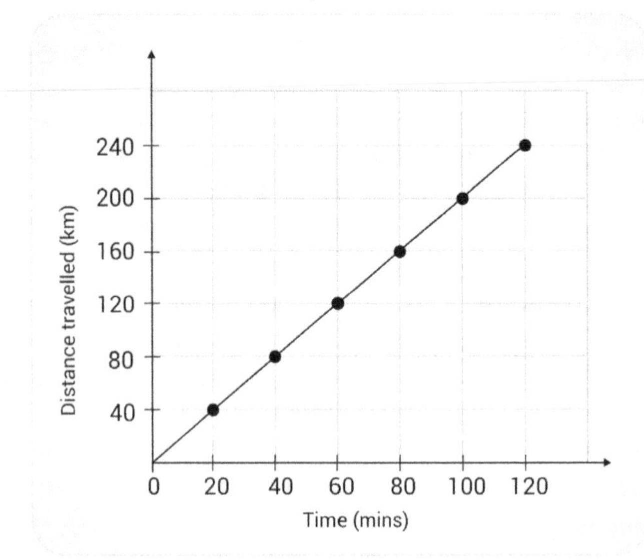

The graph above represents the relationship between distance traveled and time. To find the distance traveled in 80 minutes, the mark for 80 minutes is located at the bottom of the graph. By following this mark directly up on the graph, the corresponding point for 80 minutes is directly across from the 160-kilometer mark. This information indicates that the distance travelled in 80 minutes is 160 kilometers. To predict information not displayed on the graph, the way in which the variables change with respect to one another is determined. In this case, distance increases by 40 kilometers as time increases by 20 minutes. This information can be used to continue the data in the graph or convert the values to a table.

Determining the Equation of a Line

An equation is called **linear** if it can take the form of the equation $y = ax + b$, for any two numbers a and b. A linear equation forms a straight line when graphed on the coordinate plane. An example of a linear equation is shown below on the graph.

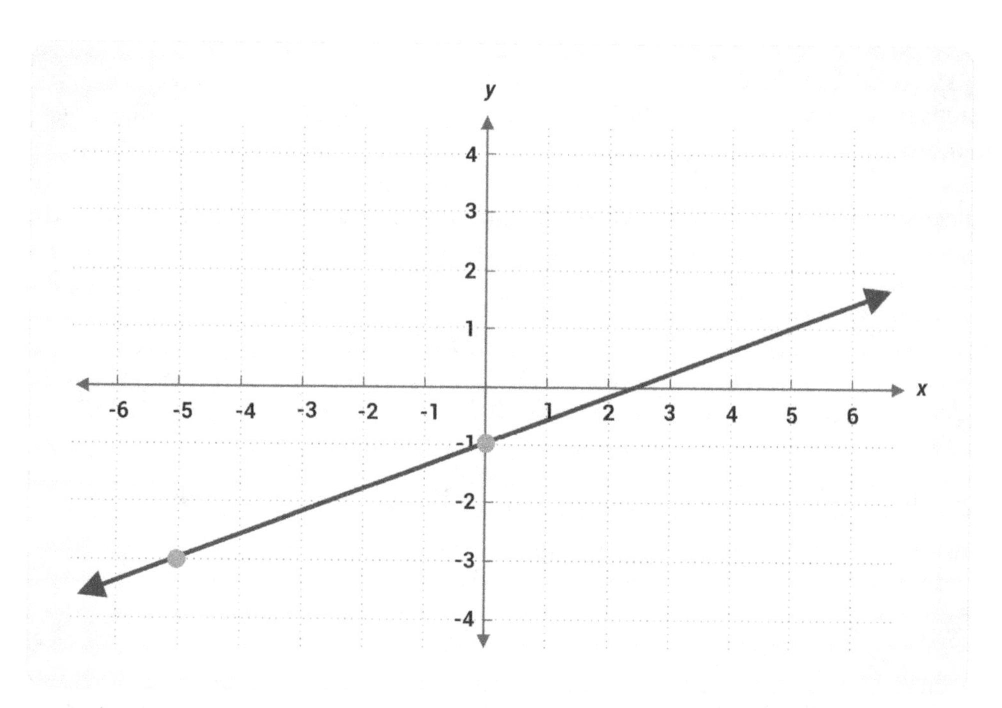

This is a graph of the following function: $y = \frac{2}{5}x - 1$. A table of values that satisfies this function is shown below.

x	y
-5	-3
0	-1
5	1
10	3

These points can be found on the graph using the form (x,y).

Given the graph of a line, its equation can be written in two ways. If the *y*-intercept is easily identified (is an integer), it and another point can be used to determine the slope. When determining:

$$\frac{change\ in\ y}{change\ in\ x}$$

from one point to another on the graph, the distance for $\frac{rise}{run}$ is being figured. The equation should be written in slope-intercept form:

$$y = mx + b$$

with m representing the slope and b representing the y-intercept.

The equation of a line can also be written by identifying two points on the graph of the line. To do so, the slope is calculated and then the values are substituted for the slope and either of the ordered pairs into the point-slope form of an equation.

The Basic Forms of Linear Equations

Linear functions may take forms other than $y = ax + b$. The most common forms of linear equations are explained below:

1. Standard Form: $Ax + By = C$, in which the slope is given by $m = \frac{-A}{B}$, and the y-intercept is given by $\frac{C}{B}$.

2. Slope-Intercept Form: $y = mx + b$, where the slope is m and the y intercept is b.

3. Point-Slope Form: $y - y_1 = m(x - x_1)$, where the slope is m and (x_1, y_1) is any point on the chosen line.

4. Two-Point Form: $\frac{y-y_1}{x-x_1} = \frac{y_2-y_1}{x_2-x_1}$, where (x_1, y_1) and (x_2, y_2) are any two distinct points on the chosen line. Note that the slope is given by $m = \frac{y_2-y_1}{x_2-x_1}$.

5. Intercept Form: $\frac{x}{x_1} + \frac{y}{y_1} = 1$, in which x_1 is the x-intercept and y_1 is the y-intercept.

These five ways to write linear equations are all useful in different circumstances. Depending on the given information, it may be easier to write one of the forms over another.

If $y = mx$, y is directly proportional to x. In this case, changing x by a factor changes y by that same factor. If $y = \frac{m}{x}$, y is inversely proportional to x. For example, if x is increased by a factor of 3, then y will be decreased by the same factor, 3.

Solving Linear Equations and Inequalities

Imagine the following problem: The sum of a number and 5 is equal to -8 times the number.

To find this unknown number, a simple equation can be written to represent the problem. Key words such as difference, equal, and times are used to form the following equation with one variable:

$$n + 5 = -8n$$

When solving for n, opposite operations are used. First, n is subtracted from $-8n$ across the equals sign, resulting in:

$$5 = -9n$$

Then, -9 is divided on both sides, leaving $n = -\frac{5}{9}$. This solution can be graphed on the number line with a dot as shown below:

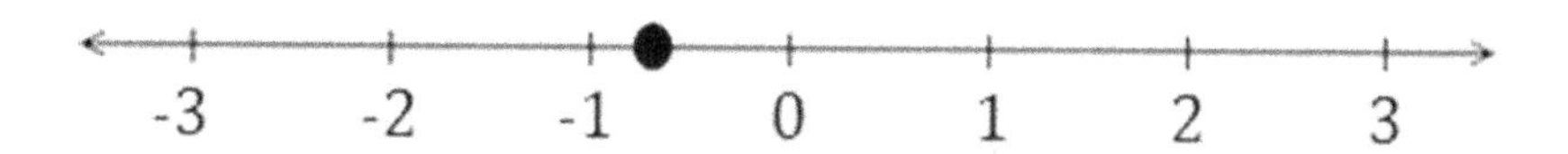

If the problem were changed to say, "The sum of a number and 5 is greater than -8 times the number," then an inequality would be used instead of an equation. Using key words again, *greater than* is represented by the symbol >. The inequality:

$$n + 5 > -8n$$

can be solved using the same techniques, resulting in:

$$n < -\frac{5}{9}$$

The only time solving an inequality differs from solving an equation is when a negative number is either multiplied by or divided by each side of the inequality. The sign must be switched in this case. For this example, the graph of the solution changes to the following graph because the solution represents all real numbers less than $-\frac{5}{9}$. Not included in this solution is $-\frac{5}{9}$ because it is a *less than* symbol, not *equal to*.

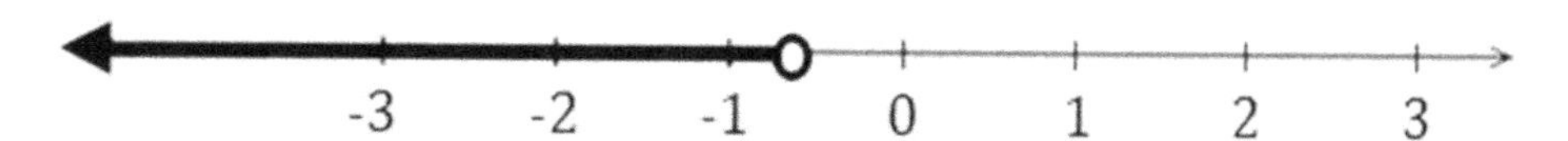

A graph of the solution set for a linear inequality shows the ordered pairs that make the statement true. The graph consists of a boundary line dividing the coordinate plane and shading on one side of the boundary. The boundary line should be graphed just as a linear equation would be graphed. If the inequality symbol is $>$ or $<$, a dashed line can be used to indicate that the line is not part of the solution set. If the inequality symbol is $\geq$ or $\leq$, a solid line can be used to indicate that the boundary line is included in the solution set. An ordered pair (*x*, *y*) on either side of the line should be chosen to test in the inequality statement. If substituting the values for *x* and *y* results in a true statement:

$$15(3) + 25(2) > 90$$

that ordered pair and all others on that side of the boundary line are part of the solution set. To indicate this, that region of the graph should be shaded. If substituting the ordered pair results in a false statement, the ordered pair and all others on that side are not part of the solution set.

Therefore, the other region of the graph contains the solutions and should be shaded.

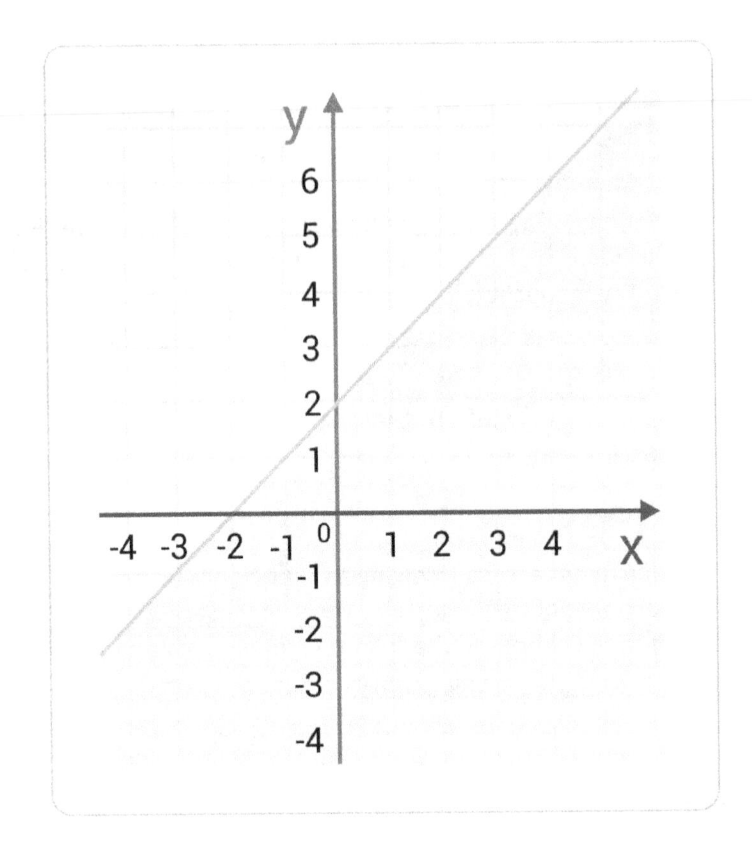

A question may simply ask whether a given ordered pair is a solution to a given inequality. To determine this, the values should be substituted for the ordered pair into the inequality. If the result is a true statement, the ordered pair is a solution; if the result is a false statement, the ordered pair is not a solution.

Solving Equations

Solving equations in one variable is the process of isolating that variable on one side of the equation. The letters in an equation are variables as they stand for unknown quantities that you are trying to solve for. The numbers attached to the variables by multiplication are called coefficients. *X* is commonly used as a variable, though any letter can be used. For example, in:

$$3x - 7 = 20$$

the variable is $3x$, and it needs to be isolated. The numbers (also called **constants**) are -7 and 20. That means $3x$ needs to be on one side of the equals sign (either side is fine), and all the numbers need to be on the other side of the equals sign.

To accomplish this, the equation must be manipulated by performing opposite operations of what already exists. Remember that addition and subtraction are opposites and that multiplication and division are opposites. Any action taken to one side of the equation must be taken on the other side to maintain equality.

So, since the 7 is being subtracted, it can be moved to the right side of the equation by adding seven to both sides:

$$3x - 7 = 20$$

$$3x - 7 + 7 = 20 + 7$$

$$3x = 27$$

Now that the variable $3x$ is on one side and the constants (now combined into one constant) are on the other side, the 3 needs to be moved to the right side. 3 and x are being multiplied together, so 3 then needs to be divided from each side.

$$\frac{3x}{3} = \frac{27}{3}$$

$$x = 9$$

Now that x has been completely isolated, we know its value.

The solution is found to be $x = 9$. This solution can be checked for accuracy by plugging $x = 9$ in the original equation. After simplifying the equation, $20 = 20$ is found, which is a true statement:

$$3 \times 9 - 7 = 20$$

$$27 - 7 = 20$$

$$20 = 20$$

Equations that require solving for a variable (**algebraic equations**) come in many forms. Here are some more examples:

Sometimes, no number is attached to the variable:

$$x + 8 = 20$$

$$x + 8 - 8 = 20 - 8$$

$$x = 12$$

A fraction may be in the variable:

$$\frac{1}{2}z + 24 = 36$$

$$\frac{1}{2}z + 24 - 24 = 36 - 24$$

$$\frac{1}{2}z = 12$$

Now we multiply the fraction by its inverse:

$$\frac{2}{1} \times \frac{1}{2} z = 12 \times \frac{2}{1}$$

$$z = 24$$

Some algebraic equations contain multiple instances of x:

$$14x + x - 4 = 3x + 2$$

All instances of x can be combined.

$$15x - 4 = 3x + 2$$

$$15x - 4 + 4 = 3x + 2 + 4$$

$$15x = 3x + 6$$

$$15x - 3x = 3x + 6 - 3x$$

$$12x = 6$$

$$\frac{12x}{12} = \frac{6}{12}$$

$$x = \frac{1}{2}$$

Number and Shape Patterns

Patterns within a sequence can come in 2 distinct forms: the items (shapes, numbers, etc.) either repeat in a constant order, or the items change from one step to another in some consistent way. The core is the smallest unit, or number of items, that repeats in a repeating pattern. For example, the pattern oo▲oo▲o... has a core that is oo▲. Knowing only the core, the pattern can be extended. Knowing the number of steps in the core allows the identification of an item in each step without drawing/writing the entire pattern out. For example, suppose the tenth item in the previous pattern must be determined. Because the core consists of three items (oo▲), the core repeats in multiples of 3. In other words, steps 3, 6, 9, 12, etc. will be ▲ completing the core with the core starting over on the next step. For the above example, the 9th step will be ▲ and the 10th will be o.

The most common patterns in which each item changes from one step to the next are arithmetic and geometric sequences. An arithmetic sequence is one in which the items increase or decrease by a constant difference. In other words, the same thing is added or subtracted to each item or step to produce the next. To determine if a sequence is arithmetic, determine what must be added or subtracted to step one to produce step two. Then, check if the same thing is added/subtracted to step two to produce step three. The same thing must be added/subtracted to step three to produce step four, and so on. Consider the pattern 13, 10, 7, 4 . . . To get from step one (13) to step two (10) by adding or subtracting requires subtracting by 3. The next step is checking if subtracting 3 from step two (10) will produce step three (7) and subtracting 3 from step three (7) will produce step four (4). In this case, the pattern holds true. Therefore, this is an arithmetic sequence in which each step is produced by subtracting 3 from the previous step. To extend the sequence, 3 is subtracted from the last step to produce the next. The next three numbers in the sequence are 1, -2, -5.

A geometric sequence is one in which each step is produced by multiplying or dividing the previous step by the same number. To determine if a sequence is geometric, decide what step one must be multiplied or divided by to produce step two. Then check if multiplying or dividing step two by the same number produces step three, and so on. Consider the pattern 2, 8, 32, 128 . . . To get from step one (2) to step two (8) requires multiplication by 4. The next step determines if multiplying step two (8) by 4 produces step three (32) and multiplying step three (32) by 4 produces step four (128). In this case, the pattern holds true. Therefore, this is a geometric sequence in which each step is produced by multiplying the previous step by 4. To extend the sequence, the last step is multiplied by 4 and repeated. The next three numbers in the sequence are 512; 2,048; 8,192.

Although arithmetic and geometric sequences typically use numbers, these sequences can also be represented by shapes. For example, an arithmetic sequence could consist of shapes with three sides, four sides, and five sides (add one side to the previous step to produce the next). A geometric sequence could consist of eight blocks, four blocks, and two blocks (each step is produced by dividing the number of blocks in the previous step by 2).

Conjectures, Predictions, or Generalizations Based on Patterns

An arithmetic or geometric sequence can be written as a formula and used to determine unknown steps without writing out the entire sequence. (Note that a similar process for repeating patterns is covered in the previous section.) An arithmetic sequence progresses by a **common difference**. To determine the common difference, any step is subtracted by the step that precedes it. In the sequence 4, 9, 14, 19 . . . the common difference, or *d*, is 5. By expressing each step as a_1, a_2, a_3, etc., a formula can be written to represent the sequence. a_1 is the first step. To produce step two, step 1 (a_1) is added to the common difference (*d*): $a_2 = a_1 + d$. To produce step three, the common difference (*d*) is added twice to a_1:

$$a_3 = a_1 + 2d$$

To produce step four, the common difference (*d*) is added three times to a_1:

$$a_4 = a_1 + 3d$$

Following this pattern allows a general rule for arithmetic sequences to be written. For any term of the sequence (a_n), the first step (a_1) is added to the product of the common difference (*d*) and one less than the step of the term ($n - 1$):

$$a_n = a_1 + (n - 1)d$$

Suppose the 8th term (a_8) is to be found in the previous sequence. By knowing the first step (a_1) is 4 and the common difference (*d*) is 5, the formula can be used:

$$a_n = a_1 + (n - 1)d$$

$$a_8 = 4 + (7)5$$

$$a_8 = 39$$

In a geometric sequence, each step is produced by multiplying or dividing the previous step by the same number. The *common ratio*, or (*r*), can be determined by dividing any step by the previous step. In the sequence 1, 3, 9, 27 . . . the common ratio (*r*) is $3(\frac{3}{1} = 3$ or $\frac{9}{3} = 3$ or $\frac{27}{9} = 3)$. Each successive step can be expressed as a product of the first step (a_1) and the common ratio (*r*) to some power. For example:

$$a_2 = a_1 \times r$$

$$a_3 = a_1 \times r \times r \text{ or } a_3 = a_1 \times r^2$$

$$a_4 = a_1 \times r \times r \times r \text{ or } a_4 = a\_1 \times r^3$$

Following this pattern, a general rule for geometric sequences can be written. For any term of the sequence (a_n), the first step (a_1) is multiplied by the common ratio (*r*) raised to the power one less than the step of the term ($n - 1$):

$$a_n = a_1 \times r^{(n-1)}$$

Suppose for the previous sequence, the 7th term (a_7) is to be found. Knowing the first step (a_1) is one, and the common ratio (*r*) is 3, the formula can be used:

$$a_n = a_1 \times r^{(n-1)} \rightarrow a_7 = (1) \times 3^6 \rightarrow a_7 = 729$$

Corresponding Terms of Two Numerical Patterns

When given two numerical patterns, the corresponding terms should be examined to determine if a relationship exists between them. Corresponding terms between patterns are the pairs of numbers that appear in the same step of the two sequences. Consider the following patterns 1, 2, 3, 4 . . . and 3, 6, 9, 12 . . . The corresponding terms are: 1 and 3; 2 and 6; 3 and 9; and 4 and 12. To identify the relationship, each pair of corresponding terms is examined, and the possibilities of performing an operation (+, −, ×, ÷) to the term from the first sequence to produce the corresponding term in the second sequence are determined. In this case:

$1 + 2 = 3$ or $1 \times 3 = 3$

$2 + 4 = 6$ or $2 \times 3 = 6$

$3 + 6 = 9$ or $3 \times 3 = 9$

$4 + 8 = 12$ or $4 \times 3 = 12$

The consistent pattern is that the number from the first sequence multiplied by 3 equals its corresponding term in the second sequence. By assigning each sequence a label (input and output) or variable (*x* and *y*), the relationship can be written as an equation. If the first sequence represents the inputs, or *x*, and the second sequence represents the outputs, or *y*, the relationship can be expressed as:

$$y = 3x$$

Consider the following sets of numbers:

a	2	4	6	8
b	6	8	10	12

To write a rule for the relationship between the values for *a* and the values for *b*, the corresponding terms (2 and 6; 4 and 8; 6 and 10; 8 and 12) are examined. The possibilities for producing *b* from *a* are:

$2 + 4 = 6$ or $2 \times 3 = 6$

$4 + 4 = 8$ or $4 \times 2 = 8$

$6 + 4 = 10$

$8 + 4 = 12$ or $8 \times 1.5 = 12$

The consistent pattern is that adding 4 to the value of *a* produces the value of *b*. The relationship can be written as the equation $a + 4 = b$.

Geometry

Classifying Two-Dimensional Figures

A **polygon** is a closed geometric figure in a plane (flat surface) consisting of at least 3 sides formed by line segments. These are often defined as two-dimensional shapes. Common two-dimensional shapes include circles, triangles, squares, rectangles, pentagons, and hexagons. Note that a circle is a two-dimensional shape without sides.

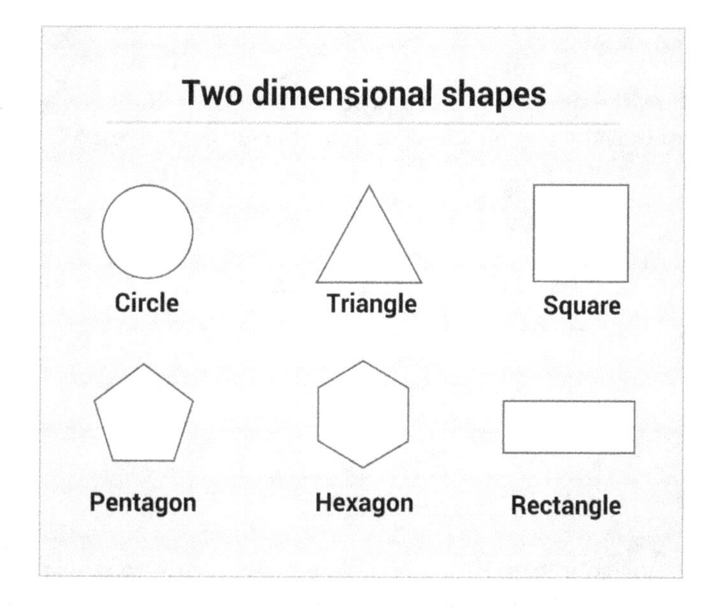

Polygons can be either convex or concave. A polygon that has interior angles all measuring less than 180° is convex. A concave polygon has one or more interior angles measuring greater than 180°. Examples are shown below.

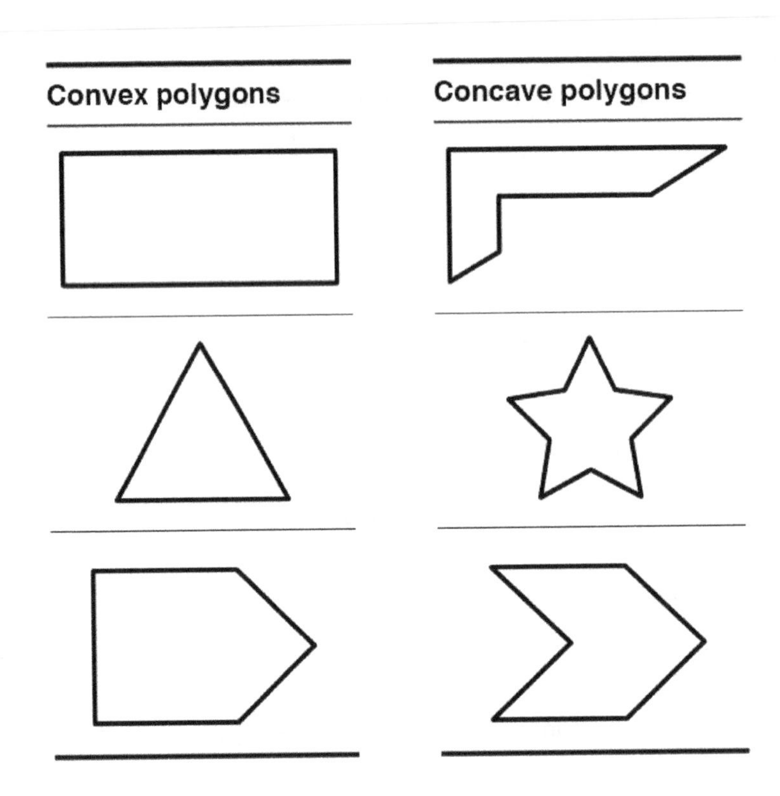

Polygons can be classified by the number of sides (also equal to the number of angles) they have. The following are the names of polygons with a given number of sides or angles:

# of Sides	Name of Polygon
3	Triangle
4	Quadrilateral
5	Pentagon
6	Hexagon
7	Septagon (or heptagon)
8	Octagon
9	Nonagon
10	Decagon

Equiangular polygons are polygons in which the measure of every interior angle is the same. The sides of equilateral polygons are always the same length. If a polygon is both equiangular and equilateral, the polygon is defined as a regular polygon.

Triangles can be further classified by their sides and angles. A triangle with its largest angle measuring 90° is a right triangle. A triangle with the largest angle less than 90° is an acute triangle. A triangle with the largest angle greater than 90° is an obtuse triangle. Below is an example of a right triangle.

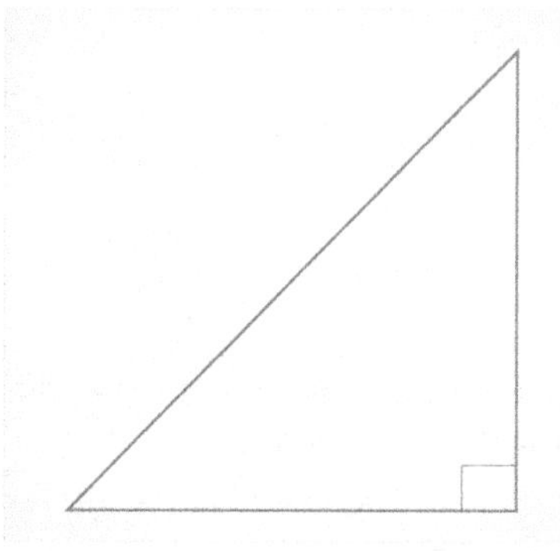

A triangle consisting of two equal sides and two equal angles is an **isosceles triangle**. A triangle with three equal sides and three equal angles is an **equilateral triangle**. A triangle with no equal sides or angles is a **scalene triangle**.

Isosceles triangle:

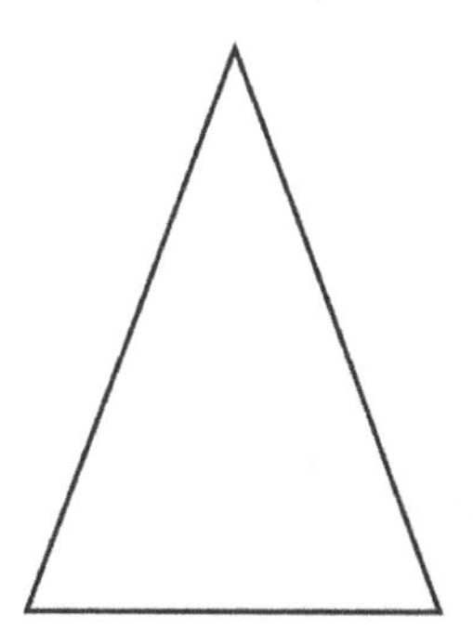

Equilateral triangle:

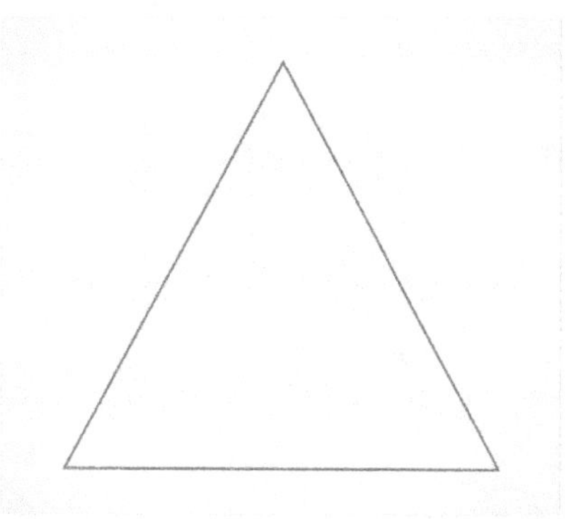

Scalene triangle:

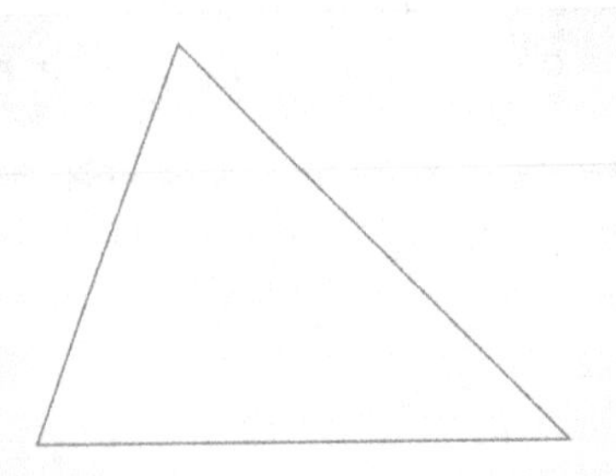

Quadrilaterals can be further classified according to their sides and angles. A quadrilateral with exactly one pair of parallel sides is called a **trapezoid**. A quadrilateral that shows both pairs of opposite sides parallel is a **parallelogram**. Parallelograms include rhombuses, rectangles, and squares. A r**hombus** has four equal sides. A **rectangle** has four equal angles (90° each). A **square** has four 90° angles and four equal sides. Therefore, a square is both a rhombus and a rectangle.

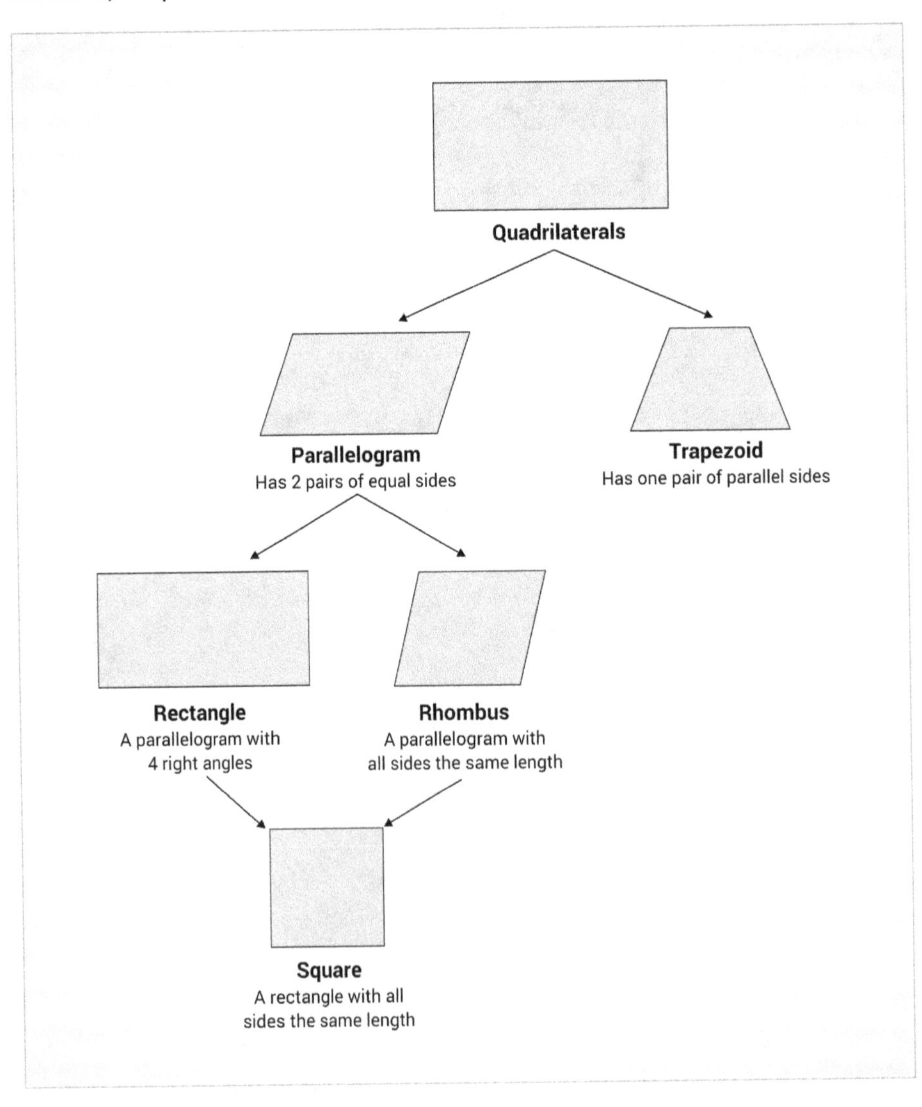

Properties of Three-Dimensional Shapes

A **solid** is a three-dimensional figure that encloses a part of space. Common three-dimensional shapes include spheres, prisms, cubes, pyramids, cylinders, and cones.

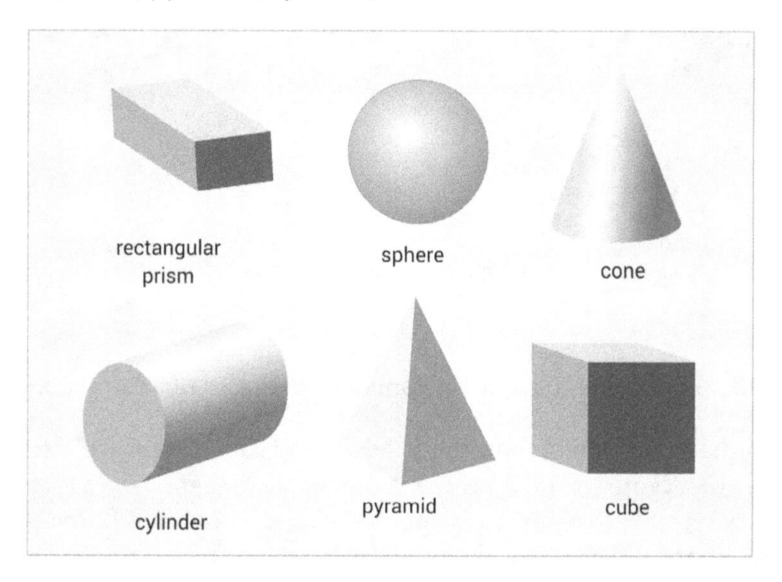

Solids consisting of all flat surfaces that are polygons are called **polyhedrons**. The two-dimensional surfaces that make up a polyhedron are called faces. Types of polyhedrons include prisms and pyramids. A **prism** consists of two parallel faces that are congruent (or the same shape and same size), and **lateral faces** going around (which are parallelograms). A prism is further classified by the shape of its base, as shown below:

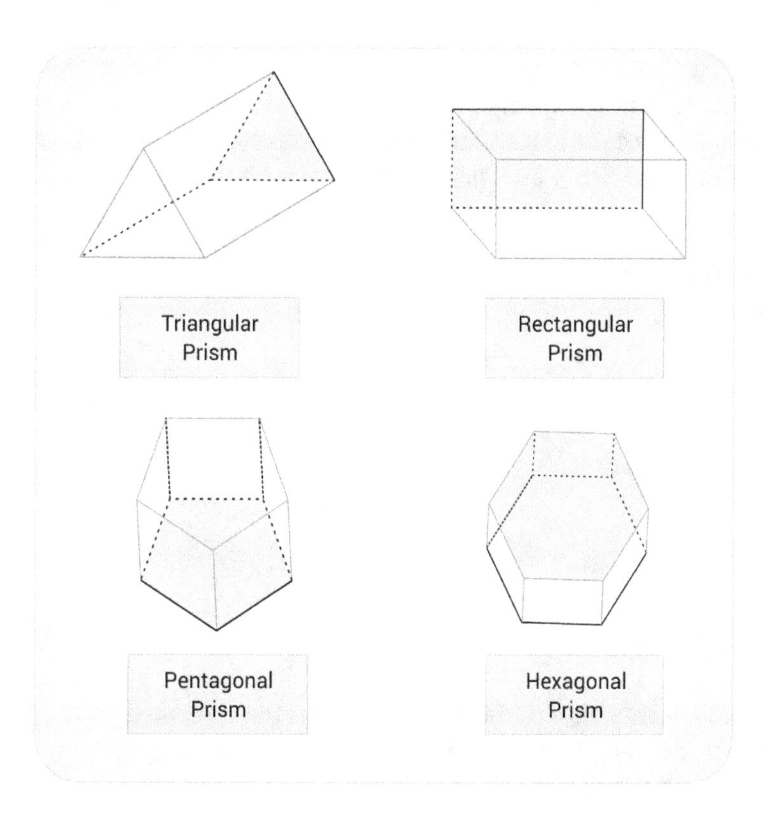

A pyramid consists of lateral faces (triangles) that meet at a common point called the vertex and one other face that is a polygon, called the base. A pyramid can be further classified by the shape of its base, as shown below.

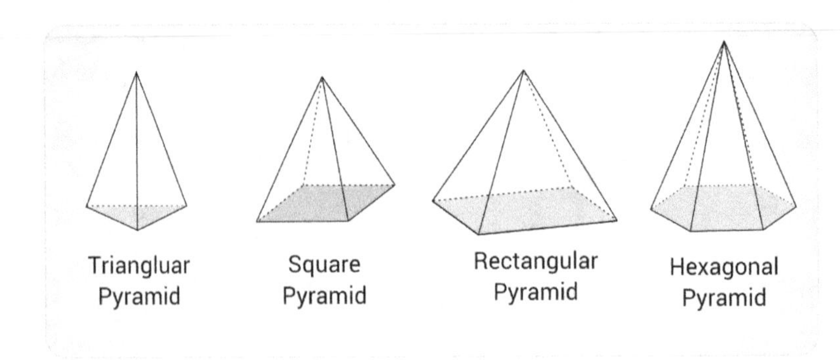

A tetrahedron is another name for a triangular pyramid. All the faces of a tetrahedron are triangles.

Solids that are not polyhedrons include spheres, cylinders, and cones. A sphere is the set of all points a given distance from a given center point. A sphere is commonly thought of as a three-dimensional circle. A cylinder consists of two parallel, congruent (same size) circles and a lateral curved surface. A cone consists of a circle as its base and a lateral curved surface that narrows to a point called the vertex.

Similar polygons are the same shape but different sizes. More specifically, their corresponding angle measures are congruent (or equal) and the length of their sides is proportional. For example, all sides of one polygon may be double the length of the sides of another. Likewise, similar solids are the same shape but different sizes. Any corresponding faces or bases of similar solids are the same polygons that are proportional by a consistent value.

Three-Dimensional Figures with Nets

A **net** is a construction of two-dimensional figures that can be folded to form a given three-dimensional figure. More than one net may exist to fold and produce the same solid, or three-dimensional figure. The bases and faces of the solid figure are analyzed to determine the polygons (two-dimensional figures) needed to form the net.

Consider the following triangular prism:

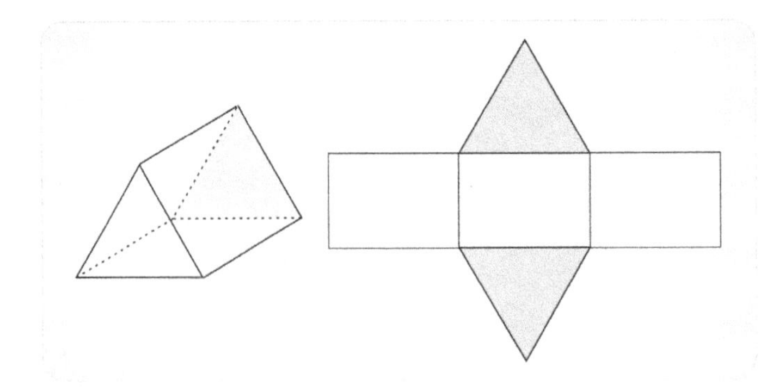

The surface of the prism consists of two triangular bases and three rectangular faces. The net beside it can be used to construct the triangular prism by first folding the triangles up to be parallel to each other, and then folding the two outside rectangles up and to the center with the outer edges touching.

Consider the following cylinder:

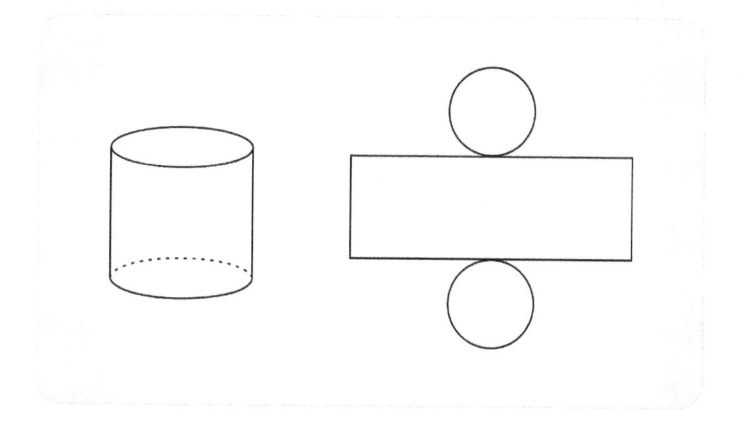

The surface consists of two circular bases and a curved lateral surface that can be opened and flattened into a rectangle. The net beside the cylinder can be used to construct the cylinder by first folding the circles up to be parallel to each other, and then curving the sides of the rectangle up to touch each other. The top and bottom of the folded rectangle should be touching the outside of both circles.

Consider the following square pyramid below on the left. The surface consists of one square base and four triangular faces. The net below on the right can be used to construct the square pyramid by folding each triangle towards the center of the square. The top points of the triangle meet at the vertex.

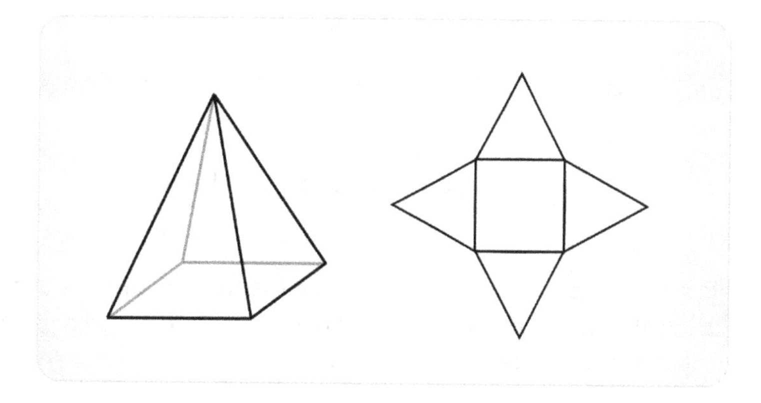

Perimeter and Area of Geometric Shapes

Perimeter is the measurement of a distance around something or the sum of all sides of a polygon. Think of perimeter as the length of the boundary, like a fence. In contrast, **area** is the space occupied by a defined enclosure, like a field enclosed by a fence.

When thinking about perimeter, think about walking around the outside of something. When thinking about area, think about the amount of space or **surface area** something takes up.

Square

The perimeter of a square is measured by adding together all of the sides. Since a square has four equal sides, its perimeter can be calculated by multiplying the length of one side by 4. Thus, the formula is $P = 4 \times s$, where s equals one side. For example, the following square has side lengths of 5 meters:

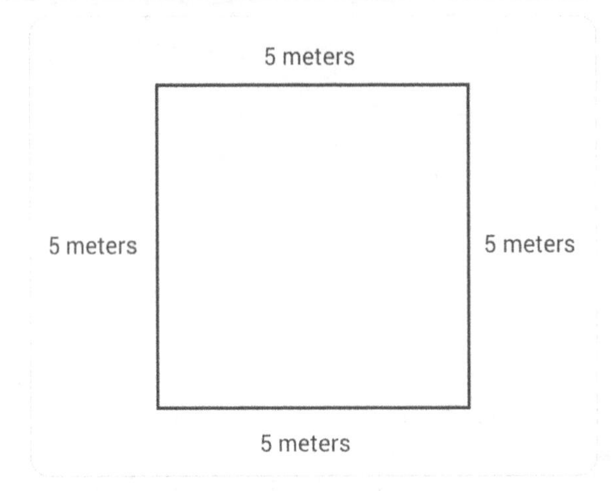

The perimeter is 20 meters because 4 times 5 is 20.

The area of a square is the length of a side squared. For example, if a side of a square is 7 centimeters, then the area is 49 square centimeters. The formula for this example is:

$$A = s^2 = 7^2 = 49 \text{ square centimeters}$$

An example is if the rectangle has a length of 6 inches and a width of 7 inches, then the area is 42 square inches:

$$A = lw = 6(7) = 42 \text{ square inches}$$

Rectangle

Like a square, a rectangle's perimeter is measured by adding together all of the sides. But as the sides are unequal, the formula is different. A rectangle has equal values for its lengths (long sides) and equal values for its widths (short sides), so the perimeter formula for a rectangle is:

$$P = l + l + w + w = 2l + 2w$$

l equals length
w equals width

The area is found by multiplying the length by the width, so the formula is $A = l \times w$.

For example, if the length of a rectangle is 10 inches and the width 8 inches, then the perimeter is 36 inches because:

$$P = 2l + 2w = 2(10) + 2(8)$$

$$20 + 16 = 36 \text{ inches}$$

Triangle

A triangle's perimeter is measured by adding together the three sides, so the formula is:

$$P = a + b + c$$

where a, b, and c are the values of the three sides. The area is the product of one-half the base and height, so the formula is:

$$A = \frac{1}{2} \times b \times h$$

It can be simplified to:

$$A = \frac{bh}{2}$$

The base is the bottom of the triangle, and the height is the distance from the base to the peak. If a problem asks to calculate the area of a triangle, it will provide the base and height.

For example, if the base of the triangle is 2 feet and the height 4 feet, then the area is 4 square feet. The following equation shows the formula used to calculate the area of the triangle:

$$\text{A} = \frac{1}{2}\text{bh} = \frac{1}{2}(2)(4) = 4 \text{ square feet}$$

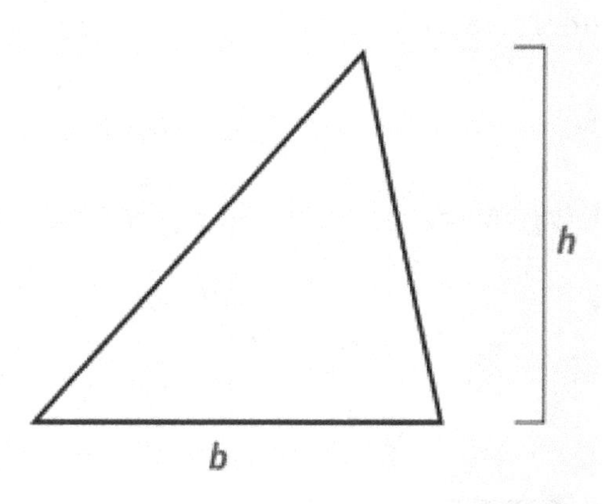

Circle

A circle's perimeter—also known as its **circumference**—is measured by multiplying the diameter by π.

Diameter is the straight line measured from a point on one side of the circle to a point directly across on the opposite side of the circle.

π is referred to as pi and is equal to 3.14 (with rounding).

So, the formula is $\pi \times d$.

This is sometimes expressed by the formula:

$$C = 2 \times \pi \times r$$

where r is the radius of the circle. These formulas are equivalent, as the radius equals half of the diameter.

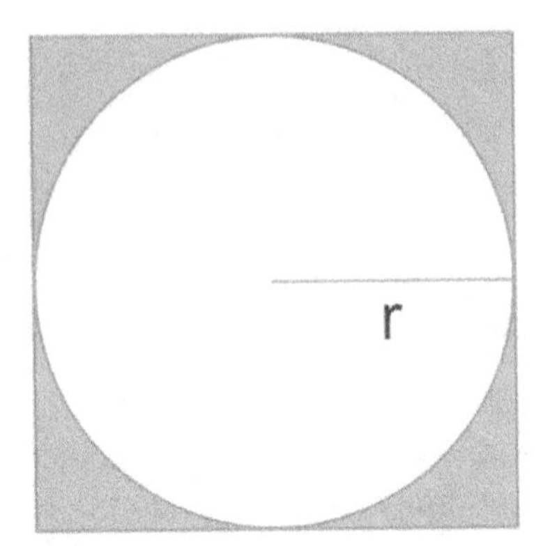

The area of a circle is calculated through the formula:

$$A = \pi \times r^2$$

The test will indicate either to leave the answer with π attached or to calculate to the nearest decimal place, which means multiplying by 3.14 for π.

Parallelogram

The perimeter of a parallelogram is measured by adding the lengths and widths together. Thus, the formula is the same as for a rectangle:

$$P = l + l + w + w = 2l + 2w$$

However, the area formula differs from the rectangle. For a parallelogram, the area is calculated by multiplying the length by the height:

$$A = h \times l$$

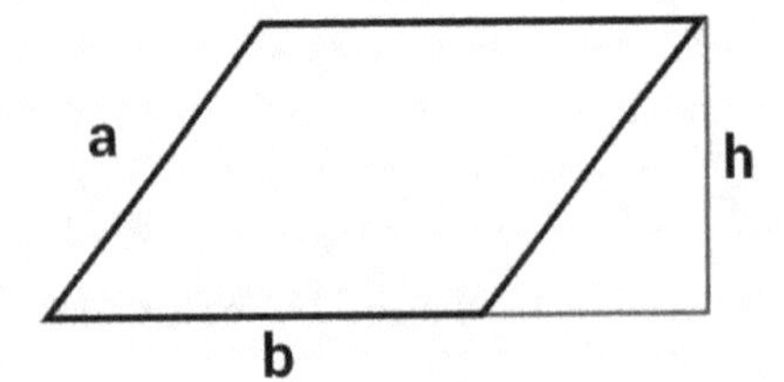

Trapezoid

The perimeter of a trapezoid is calculated by adding the two unequal bases and two equal sides, so the formula is:

$$P = a + b_1 + c + b_2$$

Although unlikely to be a test question, the formula for the area of a trapezoid is:

$$A = \frac{b_1 + b_2}{2} \times h$$

where h equals height, and b_1 and b_2 equal the bases.

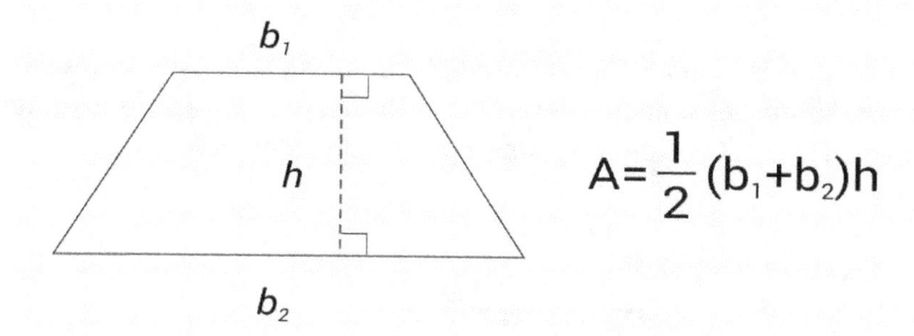

Irregular Shapes

The perimeter of an irregular polygon is found by adding the lengths of all of the sides. In cases where all of the sides are given, this will be very straightforward, as it will simply involve finding the sum of the provided lengths. Other times, a side length may be missing and must be determined before the perimeter can be calculated. Consider the example below:

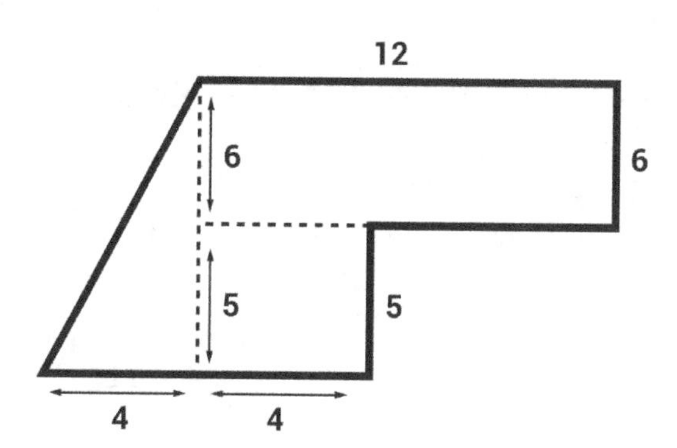

All of the side lengths are provided except for the angled side on the left. Test takers should notice that this is the hypotenuse of a right triangle. The other two sides of the triangle are provided (the base is 4 and the height is $6 + 5 = 11$). The Pythagorean Theorem can be used to find the length of the hypotenuse, remembering that $a^2 + b^2 = c^2$.

Substituting the side values provided yields:

$$(4)^2 + (11)^2 = c^2$$

Therefore:

$$c = \sqrt{16 + 121} = 11.7$$

Finally, the perimeter can be found by adding this new side length with the other provided lengths to get the total length around the figure:

$$4 + 4 + 5 + 8 + 6 + 12 + 11.7 = 50.7$$

Although units are not provided in this figure, remember that reporting units with a measurement is important.

The area of an irregular polygon is found by decomposing, or breaking apart, the figure into smaller shapes. When the area of the smaller shapes is determined, these areas are added together to produce the total area of the original figure. Consider the same example provided before:

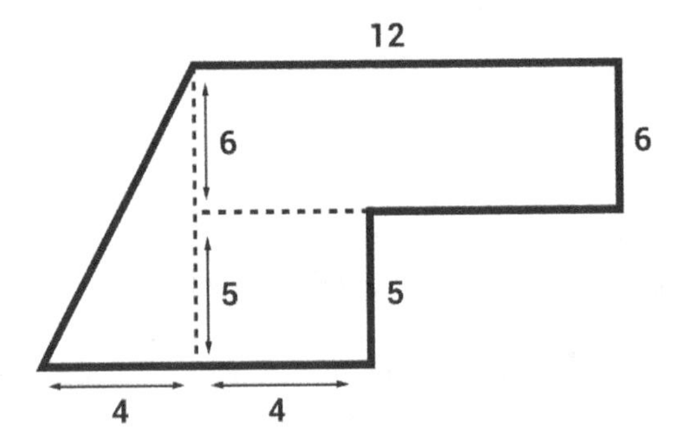

The irregular polygon is decomposed into two rectangles and a triangle. The area of the large rectangle:

$$A = l \times w \rightarrow A = 12 \times 6$$

is 72 square units. The area of the small rectangle is 20 square units ($A = 4 \times 5$). The area of the triangle:

$$A = \frac{1}{2} \times b \times h \rightarrow A = \frac{1}{2} \times 4 \times 11$$

is 22 square units. The sum of the areas of these figures produces the total area of the original polygon:

$$A = 72 + 20 + 22 \rightarrow A = 114 \text{ square units}$$

Surface Area and Volume of Geometric Shapes

The area of a two-dimensional figure refers to the number of square units needed to cover the interior region of the figure. This concept is similar to wallpaper covering the flat surface of a wall. For example, if a rectangle has an area of 8 square inches (written 8 in^2), it will take 8 squares, each with sides one inch in length, to cover the interior region of the rectangle. Note that area is measured in square units such as: square feet or ft^2; square yards or yd^2; square miles or mi^2.

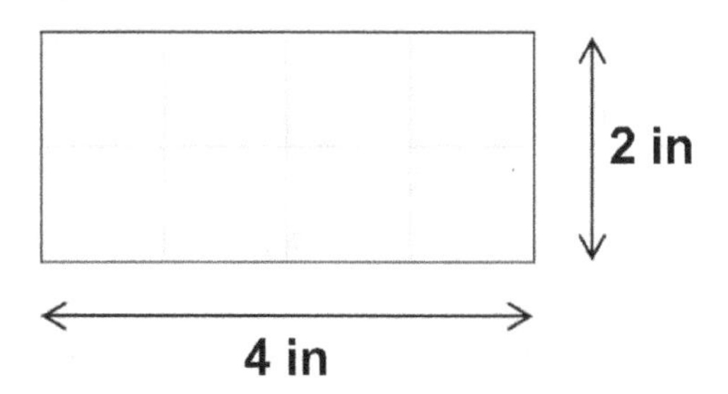

The **surface area** of a three-dimensional figure refers to the number of square units needed to cover the entire surface of the figure. This concept is similar to using wrapping paper to completely cover the outside of a box. For example, if a triangular pyramid has a surface area of 17 square inches (written $17in^2$), it will take 17 squares, each with sides one inch in length, to cover the entire surface of the pyramid. Surface area is also measured in square units.

Many three-dimensional figures (solid figures) can be represented by nets consisting of rectangles and triangles. The surface area of such solids can be determined by adding the areas of each of its faces and bases. Finding the surface area using this method requires calculating the areas of rectangles and triangles. To find the area (*A*) of a rectangle, the length (*l*) is multiplied by the width:

$$(w) \rightarrow A = l \times w$$

The area of a rectangle with a length of 8cm and a width of 4cm is calculated:

$$A = (8cm) \times (4cm) \rightarrow A = 32cm^2$$

To calculate the area (*A*) of a triangle, the product of $\frac{1}{2}$, the base (*b*), and the height (*h*) is found:

$$A = \frac{1}{2} \times b \times h$$

Note that the height of a triangle is measured from the base to the vertex opposite of it forming a right angle with the base. The area of a triangle with a base of 11cm and a height of 6cm is calculated:

$$A = \frac{1}{2} \times (11cm) \times (6cm) \rightarrow A = 33cm^2$$

Consider the following triangular prism, which is represented by a net consisting of two triangles and three rectangles.

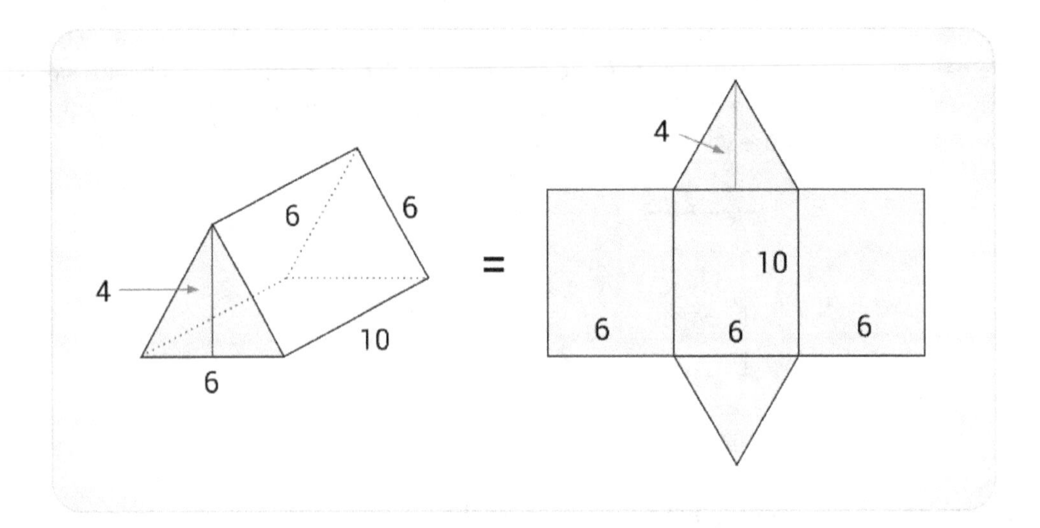

The surface area of the prism can be determined by adding the areas of each of its faces and bases. The surface area (*SA*) = area of triangle + area of triangle + area of rectangle + area of rectangle + area of rectangle.

$$SA = \left(\frac{1}{2} \times b \times h\right) + \left(\frac{1}{2} \times b \times h\right) + (l \times w) + (l \times w) + (l \times w)$$

$$SA = \left(\frac{1}{2} \times 6 \times 4\right) + \left(\frac{1}{2} \times 6 \times 4\right) + (6 \times 10) + (6 \times 10) + (6 \times 10)$$

$$SA = (12) + (12) + (60) + (60) + (60)$$

$$SA = 204 \textit{ square units}$$

Volume is the measurement of how much space an object occupies, like how much space is in the cube. Volume is useful in determining the space within a certain three-dimensional object. Volume can be calculated for a cube, rectangular prism, cylinder, pyramid, cone, and sphere. By knowing specific dimensions of the objects, the volume of the object is computed with these figures. The units for the volumes of solids can include cubic centimeters, cubic meters, cubic inches, and cubic feet.

Cube

The **cube** is the simplest figure for which volume can be determined because all dimensions in a cube are equal. In the following example, the length, width, and height of the cube are all represented by the variable *a* because these measurements are equal lengths.

The volume of any rectangular, three-dimensional object is found by multiplying its length by its width by its height. In the case of a cube, the length, width, and height are all equal lengths, represented by the variable *a*. Therefore, the equation used to calculate the volume is $(a \times a \times a)$ or a^3. In a real-world example of this situation, if the length of a side of the cube is 3 centimeters, the volume is calculated by utilizing the formula:

$$(3 \times 3 \times 3) = 27\text{cm}^3$$

Rectangular Prism

The dimensions of a **rectangular prism** are not necessarily equal as those of a cube. Therefore, the formula for a rectangular prism recognizes that the dimensions vary and use different variables to represent these lengths. The length, width, and height of a rectangular prism can be represented with the variables *a, b*, and *c*.

The equation used to calculate volume is length times width times height. Using the variables in the diagram above, this means $a \times b \times c$. In a real-world application of this situation, if *a*=2 cm, *b*=3 cm, and *c*=4 cm, the volume is calculated by utilizing the formula:

$$3 \times 4 \times 5 = 60 \text{ cm}^3$$

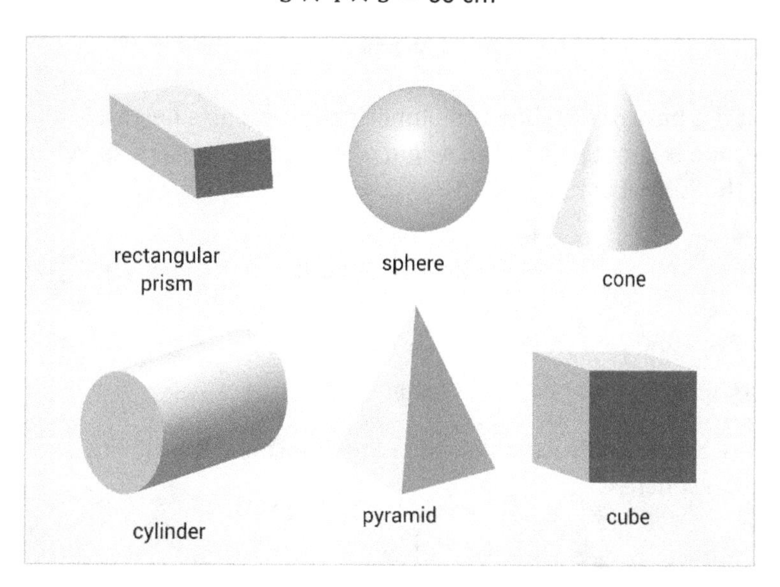

Cylinder

Discovering a cylinder's volume requires the measurement of the cylinder's base, length of the radius, and height. The height of the cylinder can be represented with variable *h*, and the radius can be represented with variable *r*.

The formula to find the volume of a cylinder is $\pi r^2 h$. Notice that πr^2 is the formula for the area of a circle. This is because the base of the cylinder is a circle. To calculate the volume of a cylinder, the slices of circles needed to build the entire height of the cylinder are added together. For example, if the radius is 5 feet and the height of the cylinder is 10 feet, the cylinder's volume is calculated by using the following equation:

$$\pi 5^2 \times 10$$

Substituting 3.14 for π, the volume is 785.4 ft^3.

Pyramid

To calculate the volume of a pyramid, the area of the base of the pyramid is multiplied by the pyramid's height by $\frac{1}{3}$. The area of the base of the pyramid is found by multiplying the base length by the base width.

Therefore, the formula to calculate a pyramid's volume is:

$$(L \times W \times H) \div 3$$

Cone

The formula to calculate the volume of a circular cone is similar to the formula for the volume of a pyramid. The primary difference in determining the area of a cone is that a circle serves as the base of a cone. Therefore, the area of a circle is used for the cone's base.

The variable *r* represents the radius, and the variable *h* represents the height of the cone. The formula used to calculate the volume of a cone is:

$$\frac{1}{3}\pi r^2 h$$

Essentially, the area of the base of the cone is multiplied by the cone's height. In a real-life example where the radius of a cone is 2 meters and the height of a cone is 5 meters, the volume of the cone is calculated by utilizing the formula:

$$\frac{1}{3}\pi 2^2 \times 5 = 21\ m^3$$

Sphere

The volume of a sphere uses π due to its circular shape.

The length of the radius, *r*, is the only variable needed to determine the sphere's volume. The formula to calculate the volume of a sphere is:

$$\frac{4}{3}\pi r^3$$

Therefore, if the radius of a sphere is 8 centimeters, the volume of the sphere is calculated by utilizing the formula:

$$\frac{4}{3}\pi(8)^3 = 2{,}144\ cm^3$$

Lines and Angles

In geometry, a **line** connects two points, has no thickness, and extends indefinitely in both directions beyond the points. If it does end at two points, it is known as a **line segment**. It is important to distinguish between a line and a line segment.

An **angle** can be visualized as a corner. It is defined as the formation of two rays connecting at a vertex that extend indefinitely. Angles are measured in degrees. Their measurement is a measure of rotation. A full rotation equals 360 degrees and represents a circle. Half of a rotation equals 180 degrees and represents a half-circle. Subsequently, 90 degrees represents a quarter-circle. Similar to the hands on a clock, an angle begins at the center point, and two lines extend indefinitely from that point in two different directions.

A clock can be useful when learning how to measure angles. At 3:00, the big hand is on the 12, and the small hand is on the 3. The angle formed is 90 degrees and is known as a **right angle**. Any angle less than 90 degrees, such as the one formed at 2:00, is known as an **acute angle**. Any angle greater than 90 degrees is known as an **obtuse angle**. The entire clock represents 360 degrees, and each clockwise increment on the clock represents an addition of 30 degrees. Therefore, 6:00 represents 180 degrees, 7:00 represents 210 degrees, etc. Angle measurement is additive. An angle can be broken into two non-overlapping angles. The total measure of the larger angle is equal to the sum of the measurements of the two smaller angles.

A **ray** is a straight path that has an endpoint on one end and extends indefinitely in the other direction. Lines are known as being **coplanar** if they are located in the same plane. Coplanar lines exist within the same two-dimensional surface. Two lines are **parallel** if they are coplanar, extend in the same direction, and never cross. They are known as being **equidistant** because they are always the same distance from each other. If lines do cross, they are known as **intersecting lines**. As discussed previously, angles are utilized throughout geometry, and their measurement can be seen through the use of an analog clock. An angle is formed when two rays begin at the same endpoint. **Adjacent angles** can be formed by forming two angles out of one shared ray. They are two side-by-side angles that also share an endpoint.

Perpendicular lines are coplanar lines that form a right angle at their point of intersection. A triangle that contains a right angle is known as a **right triangle**. The sum of the angles within any triangle is always 180 degrees. Therefore, in a right triangle, the sum of the two angles that are not right angles is 90 degrees. Any two angles that sum up to 90 degrees are known as **complementary angles**. A triangle that contains an obtuse angle is known as an **obtuse triangle**. A triangle that contains three acute angles is known as an **acute triangle**. Here is an example of a 180-degree angle, split up into an acute and obtuse angle:

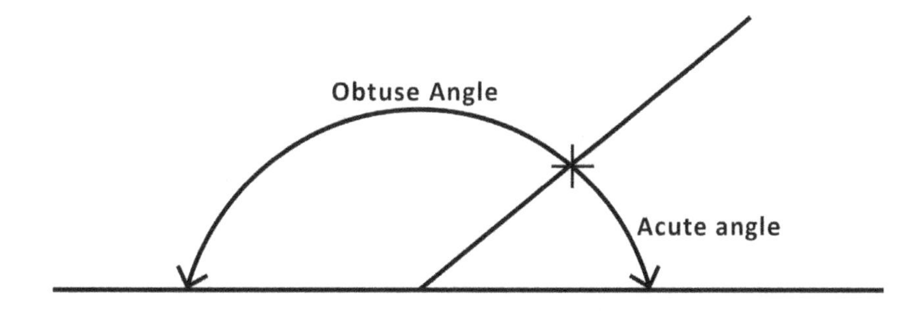

The vocabulary regarding many two-dimensional shapes is important to understand and use appropriately. Many four-sided figures can be identified using properties of angles and lines. A **quadrilateral** is a closed shape with four sides. A **parallelogram** is a specific type of quadrilateral that has two sets of parallel lines having the same length. A **trapezoid** is a quadrilateral having only one set of parallel sides. A **rectangle** is a parallelogram that has four right angles. A **rhombus** is a parallelogram with two acute angles, two obtuse angles, and four equal sides. The acute angles are of equal measure, and the obtuse angles are of equal measure. Finally, a **square** is a rhombus consisting of four right angles. It is important to note that some of these shapes share common attributes. For instance, all four-sided shapes are quadrilaterals. All squares are rectangles, but not all rectangles are squares.

Symmetry is another concept in geometry. If a two-dimensional shape can be folded along a straight line and the halves line up exactly, the figure is symmetric. The line is known as a **line of symmetry.** Circles, squares, and rectangles are examples of symmetric shapes.

Applying Angle Relationships to Solve Problems

An **angle** consists of two rays that have a common endpoint. This common endpoint is called the **vertex** of the angle. The two rays can be called **sides** of the angle. The angle below has a vertex at point *B* and the sides consist of ray *BA* and ray *BC*. An angle can be named in three ways:

1. Using the vertex and a point from each side, with the vertex letter in the middle.
2. Using only the vertex. This can only be used if it is the only angle with that vertex.
3. Using a number that is written inside the angle.

The angle below can be written $\angle ABC$ (read angle *ABC*), $\angle CBA$, $\angle B$, or $\angle 1$.

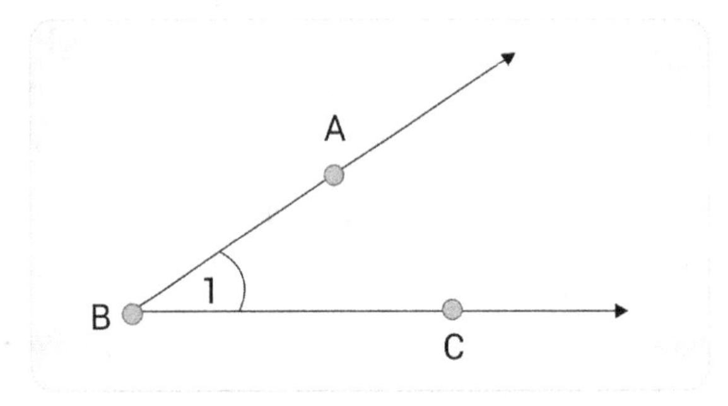

An angle divides a plane, or flat surface, into three parts: the angle itself, the interior (inside) of the angle, and the exterior (outside) of the angle. The figure below shows point *M* on the interior of the angle and point *N* on the exterior of the angle.

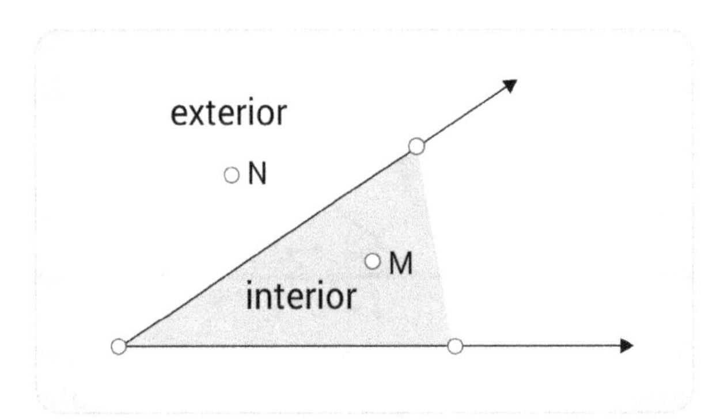

Angles can be measured in units called degrees, with the symbol °. The degree measure of an angle is between 0° and 180°, is a measure of rotation, and can be obtained by using a protractor.

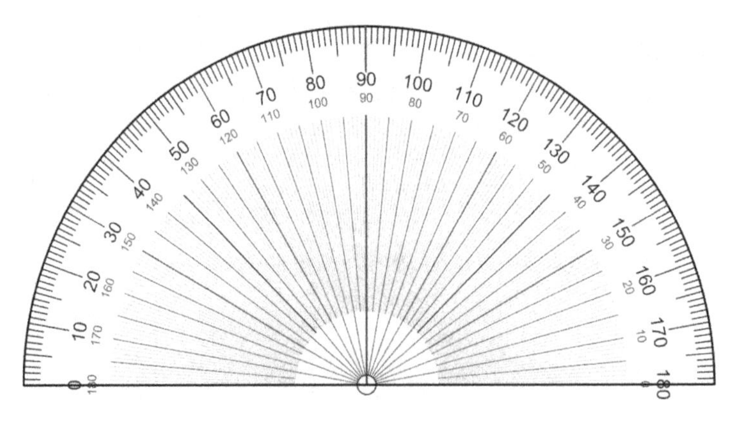

A straight angle (or simply a line) measures exactly 180°. A right angle's sides meet at the vertex to create a square corner. A right angle measures exactly 90° and is typically indicated by a box drawn in the interior of the angle. An acute angle has an interior that is narrower than a right angle. The measure of an acute angle is any value less than 90° and greater than 0°. For example, 89.9°, 47°, 12°, and 1°. An obtuse angle has an interior that is wider than a right angle. The measure of an obtuse angle is any value greater than 90° but less than 180°. For example, 90.1°, 110°, 150°, and 179.9°. Any two angles that sum up to 90 degrees are known as **complementary angles**.

- Acute angles: Less than 90°
- Obtuse angles: Greater than 90°
- Right angles: 90°
- Straight angles: 180°

To determine angle measures for adjacent angles, angles that share a common side and vertex, at least one of the angles must be known. Other information that is necessary to determine such measures include that there are 90° in a right angle, and there are 180° in a straight line. Therefore, if two adjacent angles form a right angle, they will add up to 90°, and if two adjacent angles form a straight line, they add up to 180°.

If the measurement of one of the adjacent angles is known, the other can be found by subtracting the known angle from the total number of degrees.

For example, given the following situation, if angle *a* measures 55°, find the measure of unknown angle *b*:

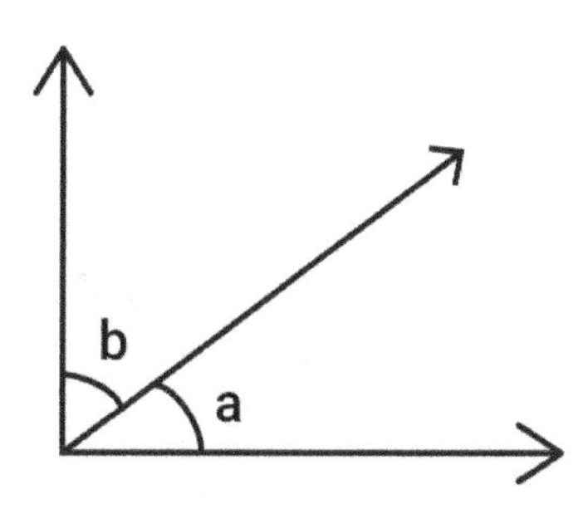

To solve this, subtract the known angle measure from 90°.

$$90° - 55° = 35°$$

The measure of *b* = 35°.

Given the following situation, if angle 1 measures 45°, find the measure of the unknown angle 2:

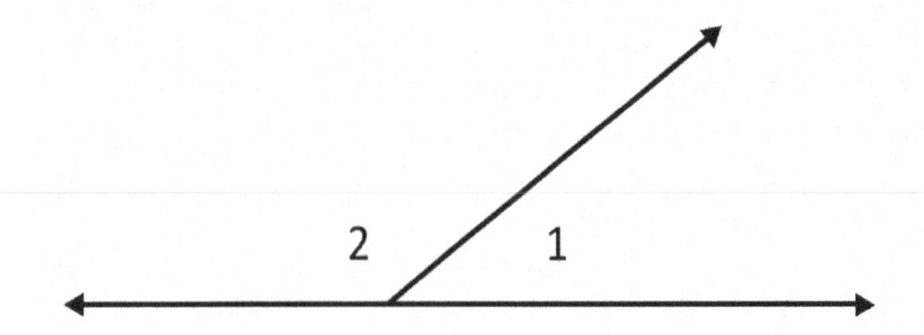

To solve this, subtract the known angle measure from 180°.

$$180° - 45° = 135°$$

The measure of angle 2 = 135°.

In the case that more than two angles are given, use the same method of subtracting the known angles from the total measure.

For example, given the following situation, if angle y = 40°, and angle z = 25°, find unknown angle x.

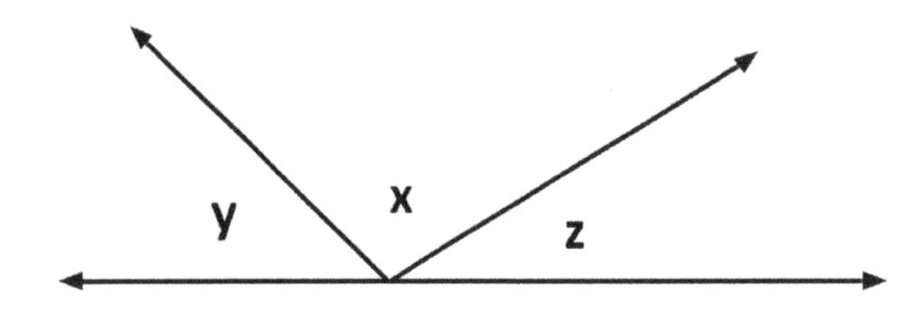

Subtract the known angles from 180°.

$$180° - 40° - 25° = 115°$$

The measure of angle x = 115°.

Measurement

Systems of Measurement

The U.S. Customary and Metric Systems of Measurement

Measurement is how an object's length, width, height, weight, and so on, are quantified. Measurement is related to counting, but it is a more refined process.

The United States customary system and the metric system each consist of distinct units to measure lengths and volume of liquids. The U.S. customary units for length, from smallest to largest, are: inch (in), foot (ft), yard (yd), and mile (mi). The metric units for length, from smallest to largest, are: millimeter (mm), centimeter (cm), decimeter (dm), meter (m), and kilometer (km). The relative size of each unit of length is shown below.

U.S. Customary	Metric	Conversion
12in = 1ft	10mm = 1cm	1in = 2.54cm
36in = 3ft = 1yd	10cm = 1dm(decimeter)	1m ≈ 3.28ft ≈ 1.09yd
5,280ft = 1,760yd = 1mi	100cm = 10dm = 1m	1mi ≈ 1.6km
	1000m = 1km	

The U.S. customary units for volume of liquids, from smallest to largest, are: fluid ounces (fl oz), cup (c), pint (pt), quart (qt), and gallon (gal). The metric units for volume of liquids, from smallest to largest, are: milliliter (mL), centiliter (cL), deciliter (dL), liter (L), and kiloliter (kL). The relative size of each unit of liquid volume is shown below.

U.S. Customary	Metric	Conversion
8fl oz = 1c	10mL = 1cL	1pt ≈ 0.473L
2c = 1pt	10cL = 1dL	1L ≈ 1.057qt
4c = 2pt = 1qt	1,000mL = 100cL = 10dL = 1L	1gal ≈ 3.785L
4qt = 1gal	1,000L = 1kL	

The U.S. customary system measures weight (how strongly Earth is pulling on an object) in the following units, from least to greatest: ounce (oz), pound (lb), and ton. The metric system measures mass (the quantity of matter within an object) in the following units, from least to greatest: milligram (mg), centigram (cg), gram (g), kilogram (kg), and metric ton (MT).

The relative sizes of each unit of weight and mass are shown below.

U.S. Measures of Weight	Metric Measures of Mass
16oz = 1lb	10mg = 1cg
2,000lb = 1 ton	100cg = 1g
	1,000g = 1kg
	1,000kg = 1MT

Note that weight and mass DO NOT measure the same thing.

Time is measured in the following units, from shortest to longest: second (sec), minute (min), hour (h), day (d), week (wk), month (mo), year (yr), decade, century, millennium. The relative sizes of each unit of time is shown below.

- 60sec = 1min
- 60min = 1h
- 24hr = 1d
- 7d = 1wk
- 52wk = 1yr
- 12mo = 1yr
- 10yr = 1 decade
- 100yrs = 1 century
- 1,000yrs = 1 millennium

Some units of measure are represented as square or cubic units depending on the solution. For example, perimeter is measured in linear units, area is measured in square units, and volume is measured in cubic units.

Also be sure to use the most appropriate unit for the thing being measured. A building's height might be measured in feet or meters while the length of a nail might be measured in inches or centimeters. Additionally, for SI units, the prefix should be chosen to provide the most succinct available value. For example, the mass of a bag of fruit would likely be measured in kilograms rather than grams or milligrams, and the length of a bacteria cell would likely be measured in micrometers rather than centimeters or kilometers.

Problems that involve measurements of length, time, volume, etc. are generally dependent upon understanding how to manipulate between various units of measurement, as well as understanding their equivalencies.

Identifying and utilizing the proper units for the scenario requires knowing how to apply the conversion rates for money, length, volume, and mass. For example, given a scenario that requires subtracting 8 inches from $2\frac{1}{2}$ feet, both values should first be expressed in the same unit (they could be expressed $\frac{2}{3}$ ft & $2\frac{1}{2}$ ft, or 8in and 30in). The desired unit for the answer may also require converting back to another unit.

Consider the following scenario:

> A parking area along the river is only wide enough to fit one row of cars and is $\frac{1}{2}$ kilometers long. The average space needed per car is 5 meters. How many cars can be parked along the river?

First, all measurements should be converted to similar units:

$$\frac{1}{2}\text{km} = 500\text{m}$$

The operation(s) needed should be identified. Because the problem asks for the number of cars, the total space should be divided by the space per car. 500 meters divided by 5 meters per car yields a total of 100 cars. Written as an expression, the meters unit cancels and the cars unit is left:

$$\frac{500m}{5m/car}$$

the same as:

$$500m \times \frac{1\ car}{5m}$$

yields 100 cars.

For an example manipulating time, Maria is scheduled to take a 90-minute test for her English class. It takes her 25 minutes to get ready and 40 minutes to ride the bus to school. If she begins to get ready at 1:10 p.m., what time will she be finished taking the test?

To find the ending time, all of the elapsed minutes must be totaled and then converted to hours.

$$25 + 40 + 90 = 155 \text{ minutes}$$

The conversion necessary for this problem is that 1 hour = 60 minutes.

The total number of minutes must be converted into hours and minutes, by dividing the total number of minutes by 60.

$$155 \div 60 = 2\ R\ 35$$

The remainder is stated as minutes. So, the total elapsed time is 2 hours and 35 minutes. If Maria begins to get ready at 1:10 p.m., 2 hours from that time is 3:10 p.m., and an additional 35 minutes would add up to 3:45 p.m. Maria can expect to be finished with everything 2 hours and 35 minutes later, at 3:45 p.m.

Converting Units of Measurement

Converting measurements in different units between the two systems can be difficult because they follow different rules. The best method is to look up an English to Metric system conversion factor and then use a series of equivalent fractions to set up an equation to convert the units of one of the measurements into those of the other. The table below lists some common conversion values that are useful for problems involving measurements with units in both systems:

English System	Metric System
1 inch	2.54 cm
1 foot	0.3048 m
1 yard	0.914 m
1 mile	1.609 km
1 ounce	28.35 g
1 pound	0.454 kg
1 fluid ounce	29.574 mL
1 quart	0.946 L
1 gallon	3.785 L

Consider the example where a scientist wants to convert 6.8 inches to centimeters. One method for converting units is to write and solve a proportion. The arrangement of values in a proportion is extremely important. The table above is used to find that there are 2.54 centimeters in every inch, so the following equation should be set up and solved:

$$\frac{6.8\ in}{1} \times \frac{2.54\ cm}{1\ in} = 17.272\ cm$$

Notice how the inches in the numerator of the initial figure and the denominator of the conversion factor cancel out. (This equation could have been written simply as:

$$6.8\ in \times 2.54\ cm = 17.272\ cm$$

but it was shown in detail to illustrate the steps). The goal in any conversion equation is to set up the fractions so that the units you are trying to convert from cancel out and the units you desire remain.

For a more complicated example, consider converting 2.15 kilograms into ounces. The first step is to convert kilograms into grams and then grams into ounces. Note that the measurement you begin with does not have to be put in a fraction.

So, in this case, 2.15 kg is by itself although it's technically the numerator of a fraction:

$$2.15\ kg \times \frac{1000g}{kg} = 2150\ g$$

Then, use the conversion factor from the table to convert grams to ounces:

$$2150g \times \frac{1\ oz}{28.35g} = 75.8\ oz$$

Now suppose that a problem requires converting 20 fluid ounces to cups. To do so, a proportion can be written using the conversion rate of 8fl oz = 1c with *x* representing the missing value. The proportion can be written in any of the following ways:

$$\frac{1}{8} = \frac{x}{20}\left(\frac{c\ for\ conversion}{fl\ oz\ for\ conversion} = \frac{unknown\ c}{fl\ oz\ given}\right); \frac{8}{1} = \frac{20}{x}\left(\frac{fl\ oz\ for\ conversion}{c\ for\ conversion} = \frac{fl\ oz\ given}{unknown\ c}\right);$$

$$\frac{1}{x} = \frac{8}{20}\left(\frac{c\ for\ conversion}{unknown\ c} = \frac{fl\ oz\ for\ conversion}{fl\ oz\ given}\right); \frac{x}{1} = \frac{20}{8}\left(\frac{unknown\ c}{c\ for\ conversion} = \frac{fl\ oz\ given}{fl\ oz\ for\ conversion}\right)$$

To solve the proportion, the ratios are cross-multiplied, and the resulting equation is solved. When cross-multiplying, all four proportions above will produce the same equation:

$$(8)(x) = (20)(1) \rightarrow 8x = 20$$

Dividing by 8 to isolate the variable *x*, the result is $x = 2.5$. The variable *x* represented the unknown number of cups. Therefore, the conclusion is that 20 fluid ounces converts (is equal) to 2.5 cups.

Sometimes converting units requires writing and solving more than one proportion. Suppose an exam question asks to determine how many hours are in 2 weeks. Without knowing the conversion rate between hours and weeks, this can be determined knowing the conversion rates between weeks and days, and between days and hours. First, weeks are converted to days, then days are converted to hours. To convert from weeks to days, the following proportion can be written:

$$\frac{7}{1} = \frac{x}{2}\left(\frac{days\ conversion}{weeks\ conversion} = \frac{days\ unknown}{weeks\ given}\right)$$

Cross-multiplying produces:

$$(7)(2) = (x)(1) \rightarrow 14 = x$$

Therefore, 2 weeks is equal to 14 days. Next, a proportion is written to convert 14 days to hours:

$$\frac{24}{1} = \frac{x}{14}\left(\frac{conversion\ hours}{conversion\ days} = \frac{unknown\ hours}{given\ days}\right)$$

Cross-multiplying produces:

$$(24)(14) = (x)(1) \rightarrow 336 = x$$

Therefore, the answer is that there are 336 hours in 2 weeks.

Measuring Lengths of Objects

The length of an object can be measured using standard tools such as rulers, yard sticks, meter sticks, and measuring tapes. The following image depicts a yardstick:

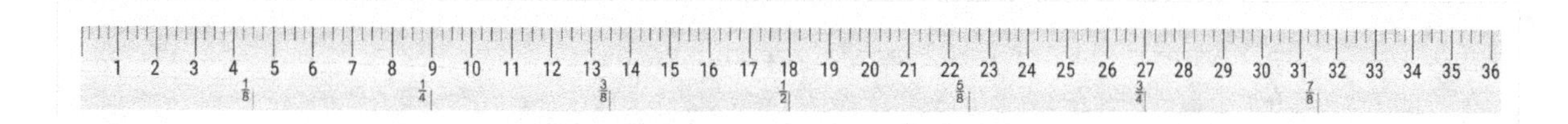

Choosing the right tool to perform the measurement requires determining whether United States customary units or metric units are desired and having a grasp of the approximate length of each unit and the approximate length of each tool. The measurement can still be performed by trial and error without the knowledge of the approximate size of the tool.

For example, if you were asked to determine the length of a room in feet, a United States customary unit, you could theoretically use a few different tools for this task. These include a ruler (typically 12 inches/1 foot long), a yardstick (3 feet/1 yard long), or a tape measure displaying feet (typically either 25 feet or 50 feet). Because the length of a room is much larger than the length of a ruler or a yardstick, a tape measure should be used to perform the measurement.

When the correct measuring tool is selected, the measurement is performed by first placing the tool directly above or below the object (if making a horizontal measurement) or directly next to the object (if making a vertical measurement). The next step is aligning the tool so that one end of the object is at the mark for zero units, then recording the unit of the mark at the other end of the object. To give the length of a paperclip in metric units, a ruler displaying centimeters is aligned with one end of the paper clip to the mark for zero centimeters.

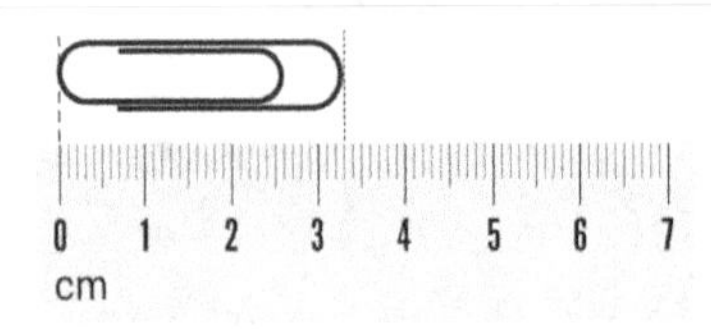

Directly down from the other end of the paperclip is the mark that measures its length. In this case, that mark is two small dashes past the 3-centimeter mark. Each small dash is 1 millimeter (or .1 centimeters). Therefore, the length of the paper clip is 3.2 centimeters.

To compare the lengths of objects, each length must be expressed with the same units. If possible, the objects should be measured with the same tool or with tools utilizing the same units. For example, a ruler and a yardstick can both measure length in inches. If the lengths of the objects are expressed in different units, these different units must be converted to the same unit before comparing them. If two lengths are expressed in the same unit, the lengths may be compared by subtracting the smaller value from the larger value. For example, suppose the lengths of two gardens are to be compared. Garden A has a length of 4 feet, and garden B has a length of 2 yards. 2 yards is converted to 6 feet so that the measurements have similar units. Then, the smaller length (4 feet) is subtracted from the larger length (6ft): 6ft – 4ft = 2ft. Therefore, garden B is 2 feet larger than garden A.

Locating Ordered Pairs in All Four Quadrants of a Rectangular Coordinate System

The **coordinate plane**, sometimes referred to as the Cartesian plane, is a two-dimensional surface consisting of a horizontal and a vertical number line. The horizontal number line is referred to as the *x*-axis, and the vertical number line is referred to as the *y*-axis. The *x*-axis and *y*-axis intersect (or cross) at a point called the origin. At the origin, the value of the *x*-axis is zero, and the value of the y-axis is zero. The coordinate plane identifies the exact location of a point that is plotted on the two-dimensional surface. Like a map, the location of all points on the plane are in relation to the origin. Along the *x*-axis (horizontal line), numbers to the right of the origin are positive and increasing in value (1,2,3, . . .) and to the left of the origin numbers are negative and decreasing in value (-1,-2,-3, . . .). Along the *y*-axis (vertical line), numbers above the origin are positive and increasing in value and numbers below the origin are negative and decreasing in value.

The *x*- and *y*-axis divide the coordinate plane into four sections. These sections are referred to as quadrant one, quadrant two, quadrant three, and quadrant four, and are often written with Roman numerals I, II, III, and IV.

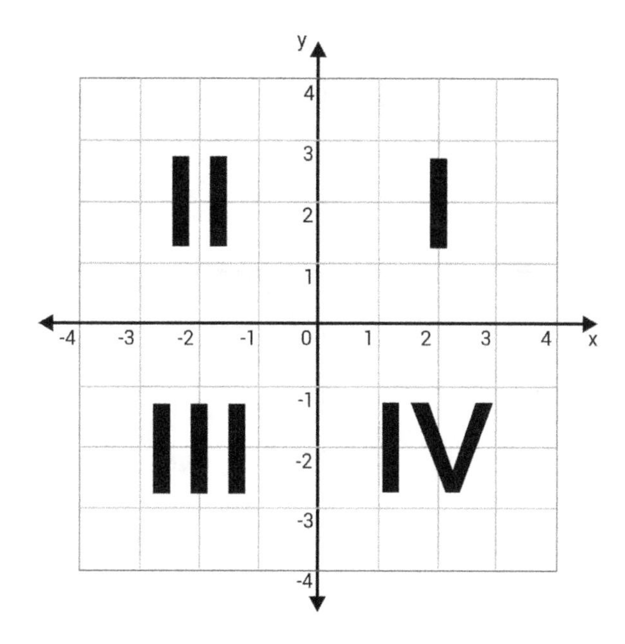

The upper right section is Quadrant I and consists of points with positive *x*-values and positive *y*-values. The upper left section is Quadrant II and consists of points with negative *x*-values and positive *y*-values. The bottom left section is Quadrant III and consists of points with negative *x*-values and negative *y*-values. The bottom right section is Quadrant IV and consists of points with positive *x*-values and negative *y*-values.

Any point within the plane can be defined by a set of **coordinates** (x, y). The coordinates consist of two numbers, x and y, which represent a position on each number line. The coordinates can also be referred to as an **ordered pair,** and (0, 0) is the ordered pair known as the **vertex**, or the origin, the point in which the axes intersect. Positive x-coordinates go to the right of the vertex, and positive y-coordinates go up. Negative x-coordinates go left, and negative y-coordinates go down.

Here is an example of the coordinate plane with a point plotted:

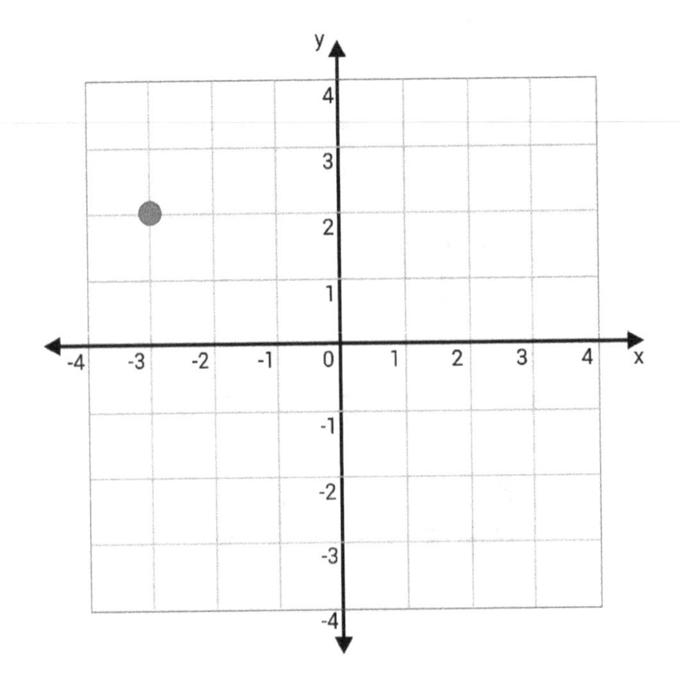

In order to plot a point on the coordinate plane, each coordinate must be considered individually. The value of x represents how many units away from the vertex the point lies on the x-axis. The value of y represents the number of units away from the vertex that the point lies on the y-axis.

The points on the coordinate plane are labeled based on their position in relation to the origin. If a point is found 4 units to the right and 2 units up from the origin, the location is described as (4, 2). These numbers are the *x*- and *y*-coordinates, always written in the order (*x*, *y*). This point is also described as lying in the first quadrant. Every point in the first quadrant has a location that is positive in the *x* and *y* directions. The following figure shows the coordinate plane with examples of points that lie in each quadrant:

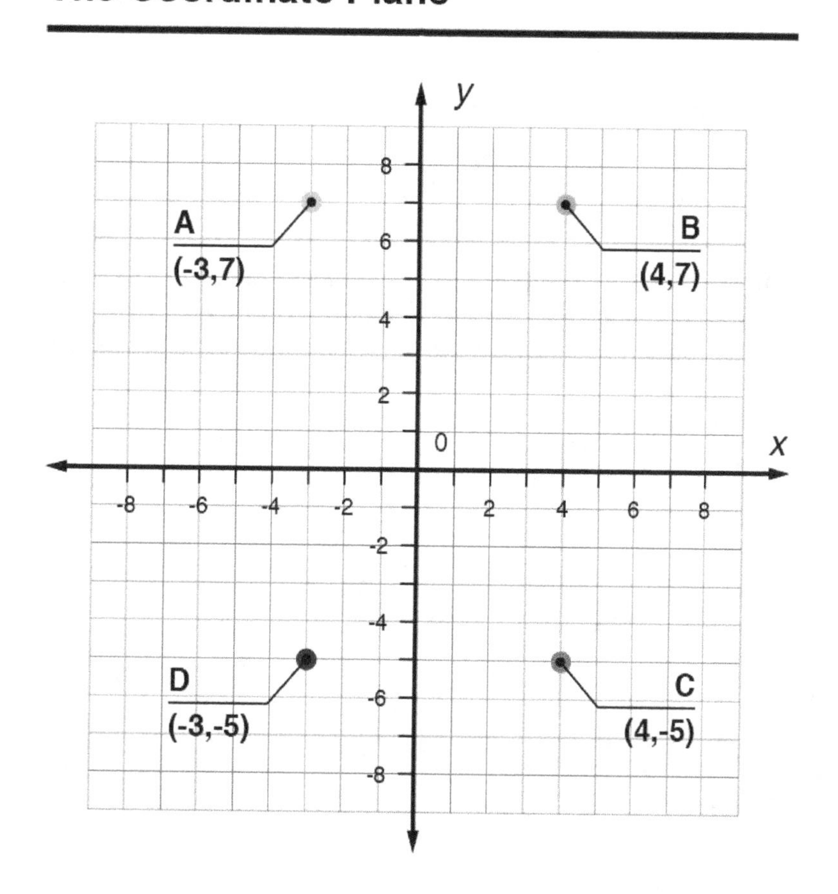

Point B lies in the first quadrant, described with positive x- and y-values, above the x-axis and to the right of the y-axis. Point A lies in the second quadrant, where the x-value is negative and the y-value is positive. This quadrant is above the x-axis and to the left of the y-axis. Point D lies in the third quadrant, where both the x- and y-values are negative. Points in this quadrant are described as being below the x-axis and to the left of the y-axis. Point C is in the fourth quadrant, where the x-value is positive and the y-value is negative.

Graphing on the Coordinate Plane Using Mathematical Problems, Tables, and Patterns

Data can be recorded using a coordinate plane. Graphs are utilized frequently in real-world applications and can be seen in many facets of everyday life. A relationship can exist between the *x*- and *y*-coordinates that are plotted on a graph, and those values can represent a set of data that can be listed in a table. Going back and forth between the table and the graph is an important concept and defining the relationship between the variables is the key that links the data to a real-life application.

For example, temperature increases during a summer day. The *x*-coordinate can be used to represent hours in the day, and the *y*-coordinate can be used to represent the temperature in degrees. The graph would show the temperature at each hour of the day. Time is almost always plotted on the *x*-axis, and utilizing different units on each axis, if necessary, is important. Labeling the axes with units is also important.

Within the first quadrant of the coordinate plane, both the x and y values are positive. Most real-world problems can be plotted in this quadrant because most real-world quantities, such as time and distance, are positive. Consider the following table of values:

X	Y
1	2
2	4
3	6
4	8

Each row gives a coordinate pair. For example, the first row gives the coordinates (1,2). Each *x*-value tells you how far to move from the origin, the point (0,0), to the right, and each *y*-value tells you how far to move up from the origin. Here is the graph of the points listed above in the table in addition to the origin:

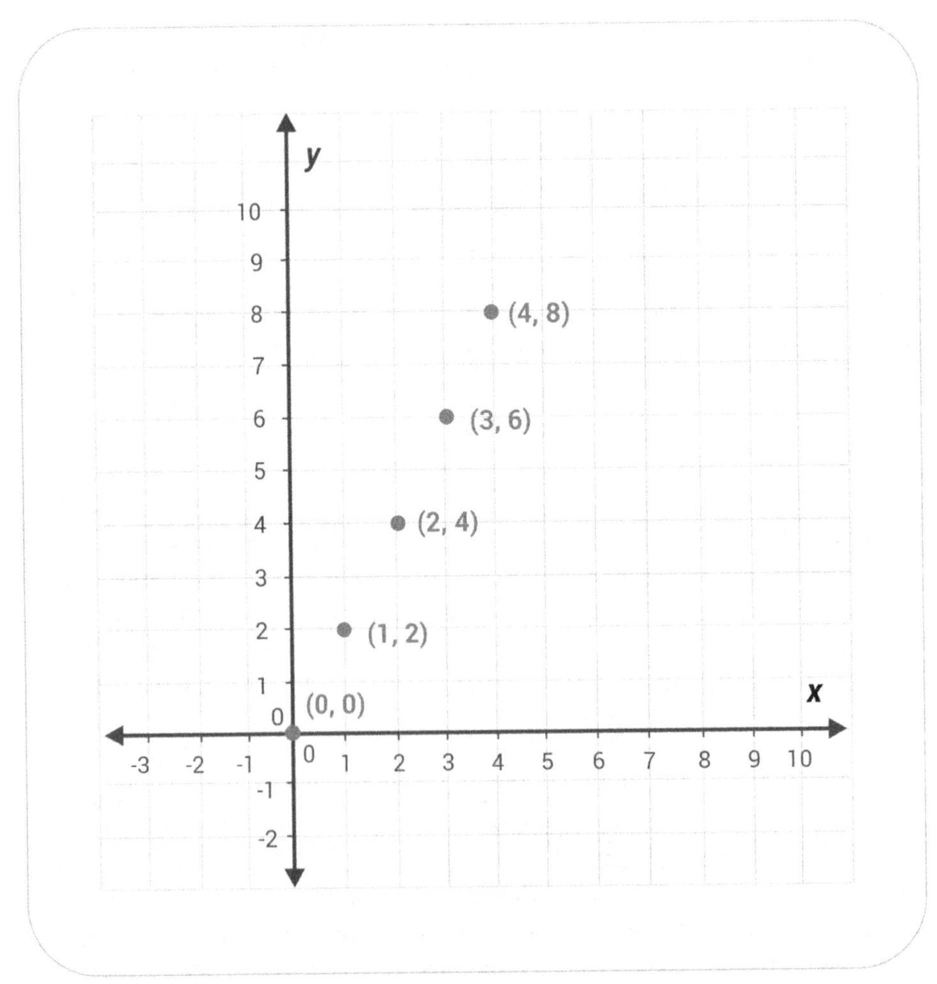

Notice that each *y*-value is found by doubling the *x*-value that forms the other portion of its coordinate pair.

Data Analysis and Probability

Graphical Representation of Data

Various graphs can be used to visually represent a given set of data. Each type of graph requires a different method of arranging data points and different calculations of the data. To construct a **histogram**, the range of the data points is divided into equal intervals. The frequency for each interval is then determined, which reveals how many points fall into each interval. A **graph** is constructed with the vertical axis representing the frequency and the horizontal axis representing the intervals. The lower value of each interval should be labeled along the horizontal axis. Finally, for each interval, a bar is drawn from the lower value of each interval to the lower value of the next interval with a height equal to the frequency of the interval. Because of the intervals, histograms do not have any gaps between bars along the horizontal axis.

To construct a **box** (or **box-and-whisker**) **plot**, the five-number summary for the data set is calculated as follows: the second quartile (Q_2) is the median of the set. The first quartile (Q_1) is the median of the values below Q_2. The third quartile (Q_3) is the median of the values above Q_2. The upper extreme is the highest value in the data set if it is not an outlier (greater than 1.5 times the interquartile range Q_3 - Q_1). The lower extreme is the least value in the data set if it is not an outlier (more than 1.5 times lower than the interquartile range). To construct the box-and-whisker plot, each value is plotted on a number line, along with any outliers. The **box** consists of Q_1 and Q_3 as its top and bottom and Q_2 as the dividing line inside the box. The **whiskers** extend from the lower extreme to Q_1 and from Q_3 to the upper extreme.

Box Plot

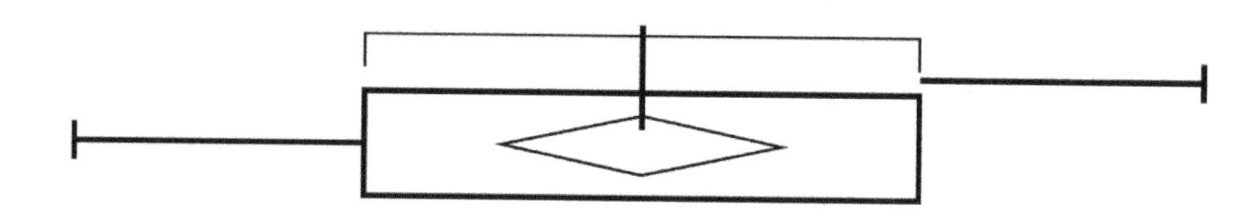

A scatter plot displays the relationship between two variables. Values for the independent variable, typically denoted by *x*, are paired with values for the dependent variable, typically denoted by *y*. Each set of corresponding values are written as an ordered pair (*x*, *y*). To construct the graph, a coordinate grid is labeled with the *x*-axis representing the independent variable and the *y*-axis representing the dependent variable. Each ordered pair is graphed.

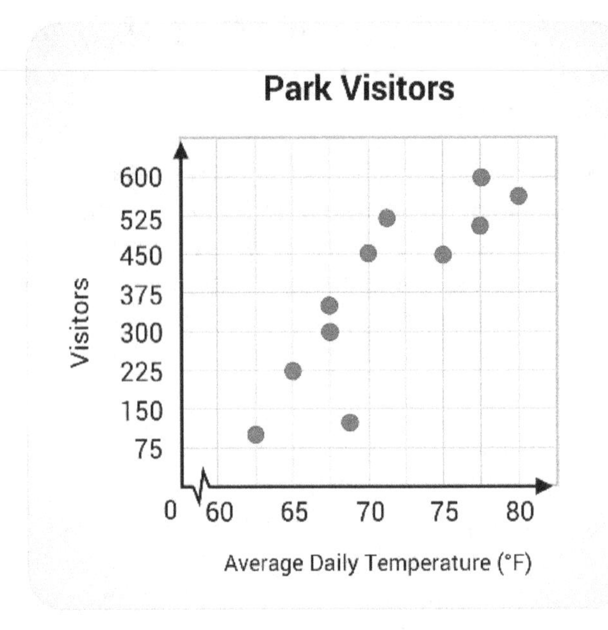

Like a scatter plot, a **line graph** compares variables that change continuously, typically over time. Paired data values (ordered pair) are plotted on a coordinate grid with the *x*- and *y*-axis representing the variables. A line is drawn from each point to the next, going from left to right. The line graph below displays cell phone use for given years (two variables) for men, women, and both sexes (three data sets).

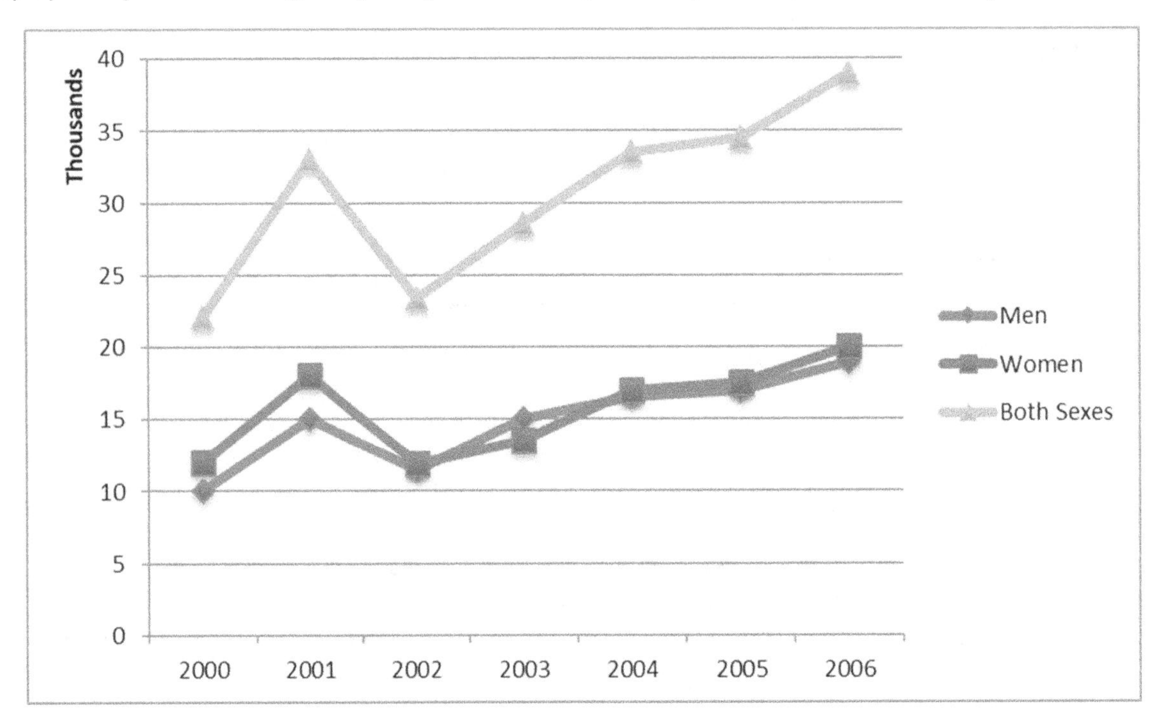

A **line plot**, also called **dot plot**, displays the frequency of data (numerical values) on a number line. To construct a line plot, a number line is used that includes all unique data values. It is marked with x's or dots above the value the number of times that the value occurs in the data set.

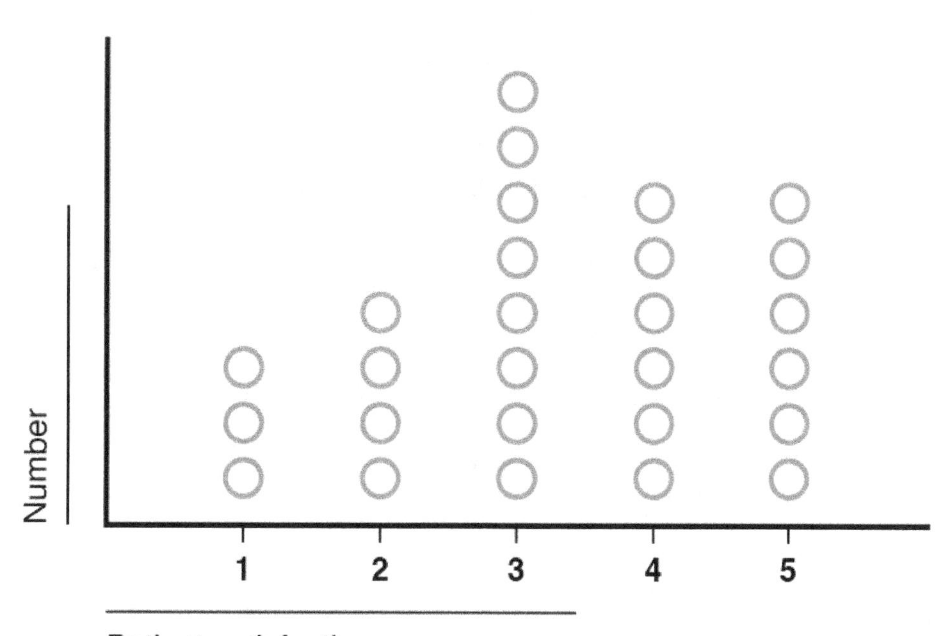

A **bar graph** is a diagram in which the quantity of items within a specific classification is represented by the height of a rectangle. Each type of classification is represented by a rectangle of equal width. Here is an example of a bar graph:

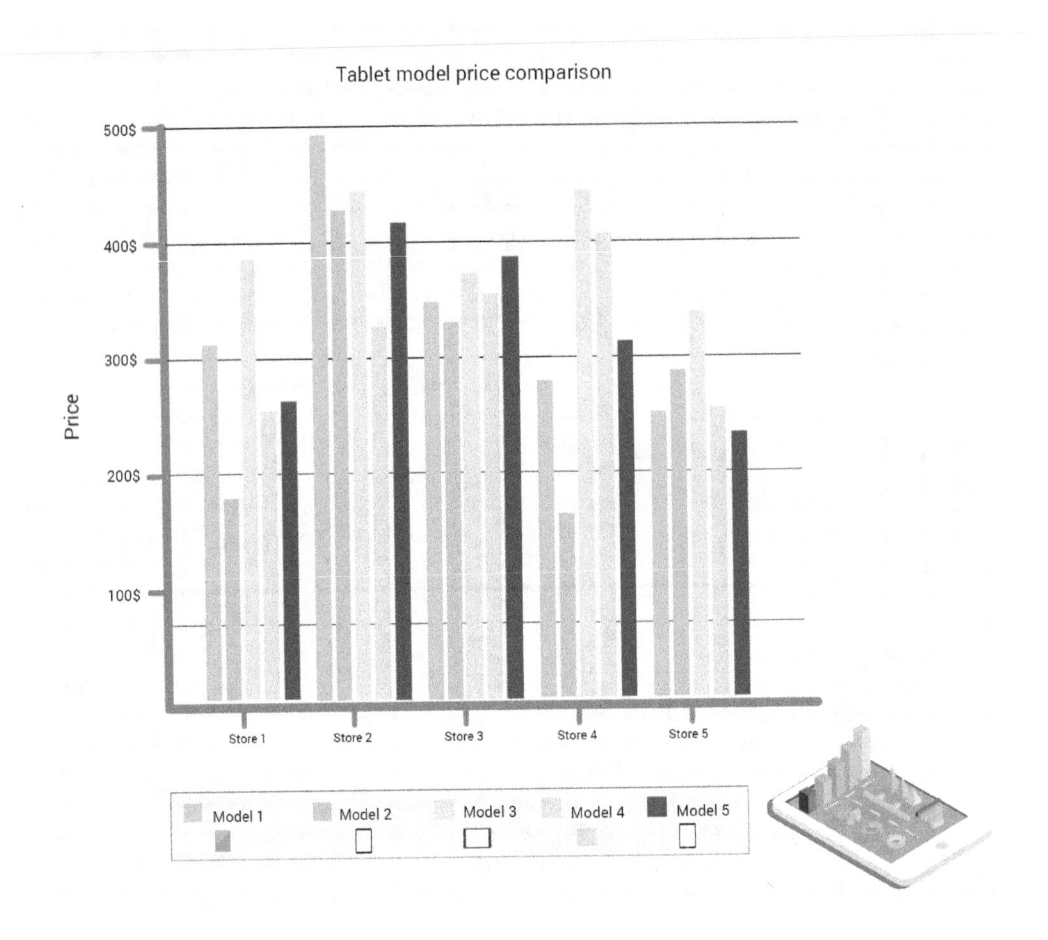

A **circle graph**, also called a **pie chart**, shows categorical data with each category representing a percentage of the whole data set. To make a circle graph, the percent of the data set for each category must be determined. To do so, the frequency of the category is divided by the total number of data points and converted to a percent. For example, if 80 people were asked what their favorite sport is and 20 responded basketball, basketball makes up 25% of the data:

$$\frac{20}{80} = 0.25 = 25\%$$

Each category in a data set is represented by a slice of the circle proportionate to its percentage of the whole.

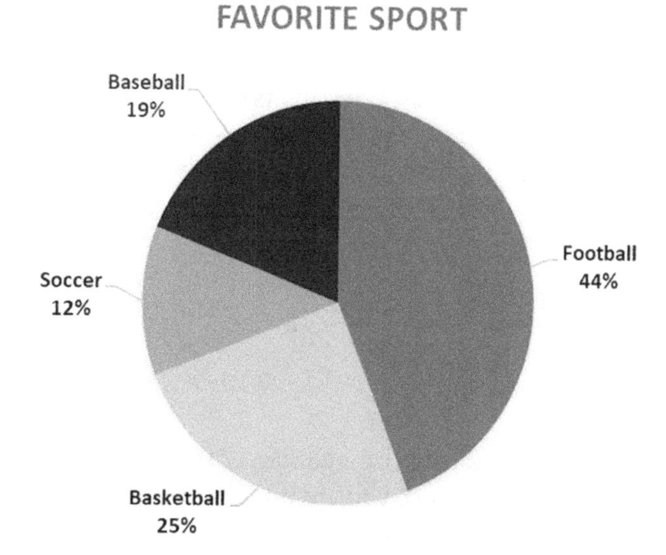

Choice of Graphs to Display Data

Choosing the appropriate graph to display a data set depends on what type of data is included in the set and what information must be displayed. Histograms and box plots can be used for data sets consisting of individual values across a wide range. Examples include test scores and incomes. Histograms and box plots will indicate the center, spread, range, and outliers of a data set. A histogram will show the shape of the data set, while a box plot will divide the set into quartiles (25% increments), allowing for comparison between a given value and the entire set.

Scatter plots and line graphs can be used to display data consisting of two variables. Examples include height and weight, or distance and time. A correlation between the variables is determined by examining the points on the graph. Line graphs are used if each value for one variable pairs with a distinct value for the other variable. Line graphs show relationships between variables.

Line plots, bar graphs, and circle graphs are all used to display categorical data, such as surveys. Line plots and bar graphs both indicate the frequency of each category within the data set. A line plot is used when the categories consist of numerical values. For example, the number of hours of TV watched by individuals is displayed on a line plot. A bar graph is used when the categories consists of words. For example, the favorite ice cream of individuals is displayed with a bar graph. A circle graph can be used to display either type of categorical data. However, unlike line plots and bar graphs, a circle graph does not indicate the frequency of each category. Instead, the circle graph represents each category as its percentage of the whole data set.

Measures of Center and Range

The center of a set of data (statistical values) can be represented by its mean, median, or mode. These are sometimes referred to as measures of central tendency. The **mean** is the average of the data set. The mean can be calculated by adding the data values and dividing by the sample size (the number of data points). Suppose a student has test scores of 93, 84, 88, 72, 91, and 77. To find the mean, or average, the scores are added, and the sum is divided by 6 because there are 6 test scores:

$$\frac{93+84+88+72+91+77}{6}=\frac{505}{6}=84.17$$

Given the mean of a data set and the sum of the data points, the sample size can be determined by dividing the sum by the mean. Suppose you are told that Kate averaged 12 points per game and scored a total of 156 points for the season. The number of games that she played (the sample size or the number of data points) can be determined by dividing the total points (sum of data points) by her average (mean of data points):

$$\frac{156}{12}=13$$

Therefore, Kate played in 13 games this season.

If given the mean of a data set and the sample size, the sum of the data points can be determined by multiplying the mean and sample size. Suppose you are told that Tom worked 6 days last week for an average of 5.5 hours per day. The total number of hours worked for the week (sum of data points) can be determined by multiplying his daily average (mean of data points) by the number of days worked (sample size):

$$5.5\times 6=33$$

Therefore, Tom worked a total of 33 hours last week.

The **median** of a data set is the value of the data point in the middle when the sample is arranged in numerical order. To find the median of a data set, the values are written in order from least to greatest. The lowest and highest values are simultaneously eliminated, repeating until the value in the middle remains. Suppose the salaries of math teachers are: $35,000; $38,500; $41,000; $42,000; $42,000; $44,500; $49,000. The values are listed from least to greatest to find the median. The lowest and highest values are eliminated until only the middle value remains. Repeating this step three times reveals a median salary of $42,000. If the sample set has an even number of data points, two values will remain after all others are eliminated. In this case, the mean of the two middle values is the median. Consider the following data set: 7, 9, 10, 13, 14, 14. Eliminating the lowest and highest values twice leaves two values, 10 and 13, in the middle. The mean of these values $\left(\frac{10+13}{2}\right)$ is the median. Therefore, the set has a median of 11.5.

The **mode** of a data set is the value that appears most often. A data set may have a single mode, multiple modes, or no mode. If different values repeat equally as often, multiple modes exist. If no value repeats, no mode exists. Consider the following data sets:

- A: 7, 9, 10, 13, 14, 14
- B: 37, 44, 33, 37, 49, 44, 51, 34, 37, 33, 44
- C: 173, 154, 151, 168, 155

Set A has a mode of 14. Set B has modes of 37 and 44. Set C has no mode.

The **range** of a data set is the difference between the highest and the lowest values in the set. The range can be considered to be the span of the data set. To determine the range, the smallest value in the set is subtracted from the largest value. The ranges for the data sets A, B, and C above are calculated as follows: A: $14 - 7 = 7$; B: $51 - 33 = 18$; C: $173 - 151 = 22$.

Best Description of a Set of Data

Measures of central tendency, namely mean, median, and mode, describe characteristics of a set of data. Specifically, they are intended to represent a typical value in the set by identifying a central position of the set. Depending on the characteristics of a specific set of data, different measures of central tendency are more indicative of a typical value in the set.

When a data set is grouped closely together with a relatively small range and the data is spread out somewhat evenly, the **mean** is an effective indicator of a typical value in the set. Consider the following data set representing the height of sixth grade boys in inches: 61 inches, 54 inches, 58 inches, 63 inches, 58 inches. The mean of the set is 58.8 inches. The data set is grouped closely (the range is only 9 inches) and the values are spread relatively evenly (three values below the mean and two values above the mean). Therefore, the mean value of 58.8 inches is an effective measure of central tendency in this case.

When a data set contains a small number of values, with one either extremely large or extremely small when compared to the other values, the mean is not an effective measure of central tendency. Consider the following data set representing annual incomes of homeowners on a given street: $71,000; $74,000; $75,000; $77,000; $340,000. The mean of this set is $127,400. This figure does not indicate a typical value in the set, which contains four out of five values between $71,000 and $77,000. The **median** is a much more effective measure of central tendency for data sets such as these. Finding the middle value diminishes the influence of **outliers**, or numbers that may appear out of place, like the $340,000 annual income. The median for this set is $75,000 which is a much more typical value in the set.

The **mode** of a data set is a useful measure of central tendency for categorical data when each piece of data is an option from a category. Consider a survey of 31 commuters asking how they get to work with results summarized below.

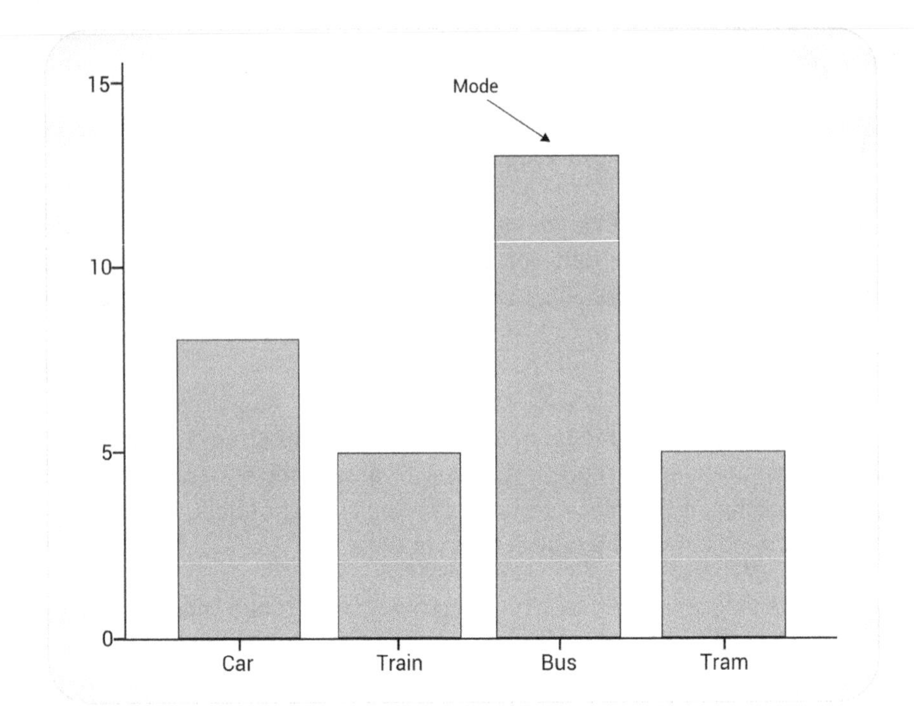

The mode for this set represents the value, or option, of the data that repeats most often. This indicates that the bus is the most popular method of transportation for the commuters.

Effects of Changes in Data

Changing all values of a data set in a consistent way produces predictable changes in the measures of the center and range of the set. A linear transformation changes the original value into the new value by either adding a given number to each value, multiplying each value by a given number, or both. Adding (or subtracting) a given value to each data point will increase (or decrease) the mean, median, and any modes by the same value. However, the range will remain the same due to the way that range is calculated. Multiplying (or dividing) a given value by each data point will increase (or decrease) the mean, median, and any modes, and the range by the same factor.

Consider the following data set, call it set P, representing the price of different cases of soda at a grocery store: $4.25, $4.40, $4.75, $4.95, $4.95, $5.15. The mean of set P is $4.74. The median is $4.85. The mode of the set is $4.95. The range is $0.90. Suppose the state passes a new tax of $0.25 on every case of soda sold. The new data set, set T, is calculated by adding $0.25 to each data point from set P. Therefore, set T consists of the following values: $4.50, $4.65, $5.00, $5.20, $5.20, $5.40. The mean of set T is $4.99. The median is $5.10. The mode of the set is $5.20. The range is $.90. The mean, median and mode of set T is equal to $0.25 added to the mean, median, and mode of set P. The range stays the same.

Now suppose, due to inflation, the store raises the cost of every item by 10 percent. Raising costs by 10 percent is calculated by multiplying each value by 1.1. The new data set, set I, is calculated by multiplying each data point from set T by 1.1. Therefore, set I consists of the following values: $4.95,

$5.12, $5.50, $5.72, $5.72, $5.94. The mean of set *I* is $5.49. The median is $5.61. The mode of the set is $5.72. The range is $0.99. The mean, median, mode, and range of set *I* is equal to 1.1 multiplied by the mean, median, mode, and range of set *T* because each increased by a factor of 10 percent.

Describing a Set of Data

A **set of data** can be described in terms of its center, spread, shape and any unusual features. The center of a data set can be measured by its mean, median, or mode. The spread of a data set refers to how far the data points are from the center (mean or median). The spread can be measured by the range or the quartiles and interquartile range. A data set with all its data points clustered around the center will have a small spread. A data set covering a wide range of values will have a large spread.

When a data set is displayed as a **histogram** or frequency distribution plot, the shape indicates if a sample is normally distributed, symmetrical, or has measures of skewness or kurtosis. When graphed, a data set with a **normal distribution** will resemble a bell curve.

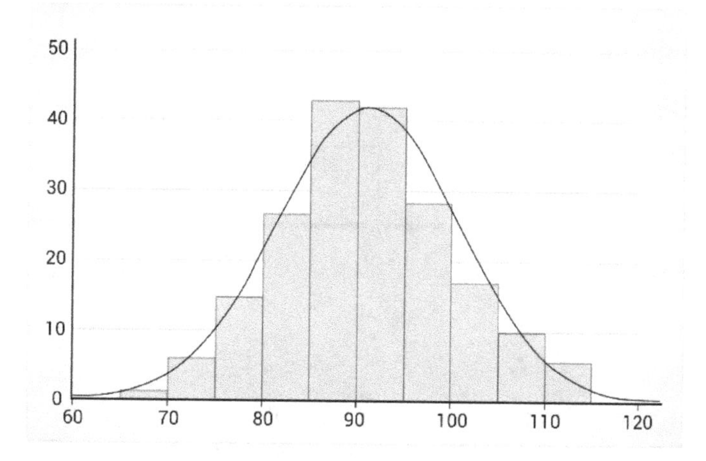

If the data set is symmetrical, each half of the graph when divided at the center is a mirror image of the other. If the graph has fewer data points to the right, the data is **skewed right**. If it has fewer data points to the left, the data is **skewed left**.

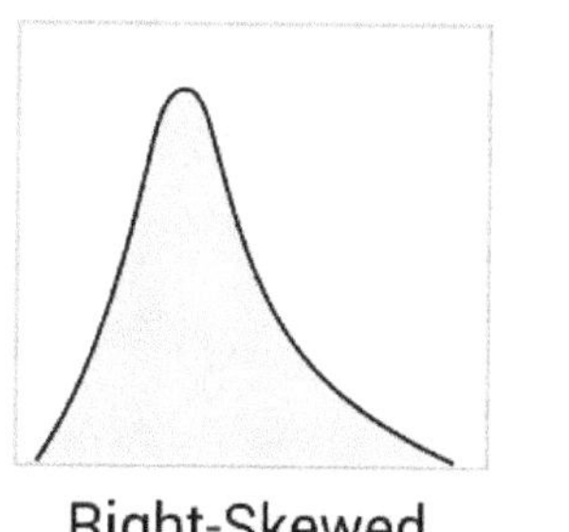
Right-Skewed

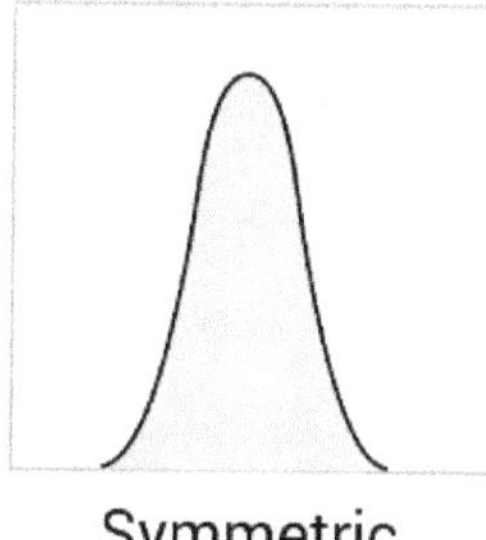
Symmetric

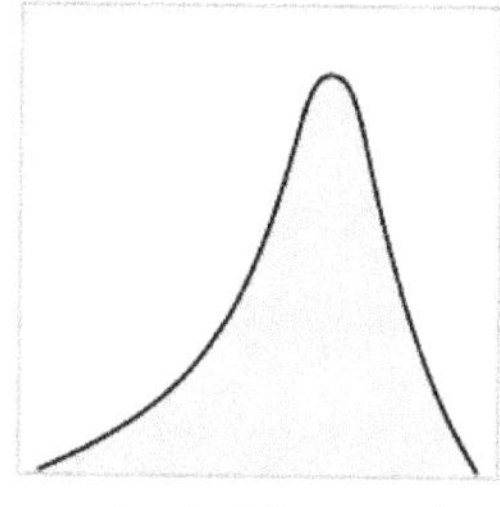
Left-Skewed

Kurtosis is a measure of whether the data is heavy-tailed with a high number of outliers, or light-tailed with a low number of outliers.

A description of a data set should include any unusual features such as gaps or outliers. A **gap** is a span within the range of the data set containing no data points. An **outlier** is a data point with a value either extremely large or extremely small when compared to the other values in the set.

Interpreting Displays of Data

A set of data can be visually displayed in various forms allowing for quick identification of characteristics of the set. **Histograms**, such as the one shown below, display the number of data points (vertical axis) that fall into given intervals (horizontal axis) across the range of the set. Suppose the histogram below displays IQ scores of students. Histograms can display the center, spread, shape, and any unusual characteristics of a data set.

Histogram

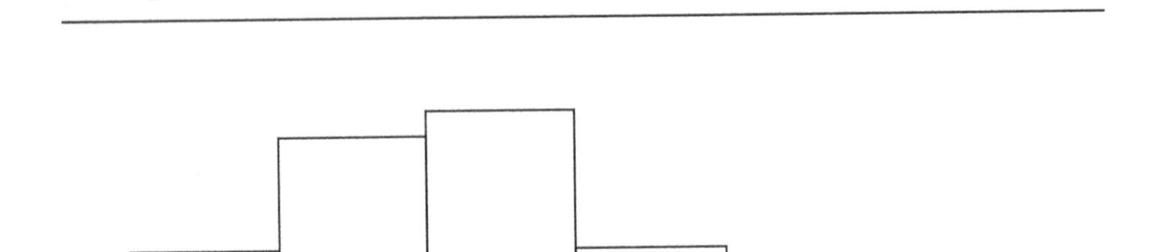

As mentioned, a **box plot**, also called a box-and-whisker plot, divides the data points into four groups and displays the five-number summary for the set, as well as any outliers. The five-number summary consists of:

- The lower extreme: the lowest value that is not an outlier
- The higher extreme: the highest value that is not an outlier
- The median of the set: also referred to as the second quartile or Q_2
- The first quartile or Q_1: the median of values below Q_2
- The third quartile or Q_3: the median of values above Q_2

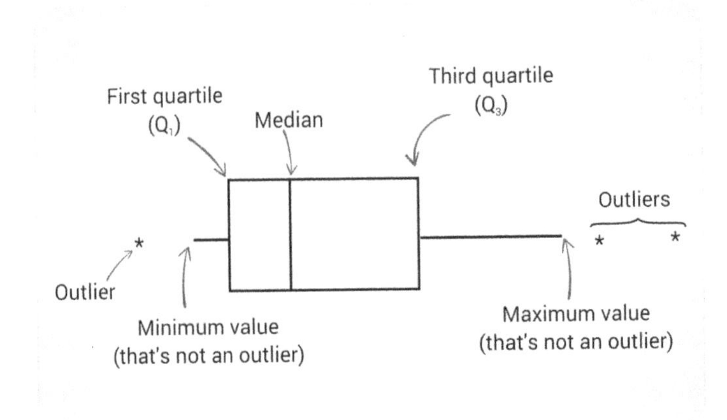

Suppose the box plot displays IQ scores for 12th grade students at a given school. The five-number summary of the data consists of: lower extreme (67); upper extreme (127); Q_2 or median (100); Q_1 (91); Q_3 (108); and outliers (135 and 140). Although all data points are not known from the plot, the points are divided into four quartiles each, including 25% of the data points. Therefore, 25% of students scored between 67 and 91, 25% scored between 91 and 100, 25% scored between 100 and 108, and 25% scored between 108 and 127. These percentages include the normal values for the set and exclude the outliers. This information is useful when comparing a given score with the rest of the scores in the set.

A **scatter plot** is a mathematical diagram that visually displays the relationship or connection between two variables. The independent variable is placed on the *x*-axis (the horizontal axis), and the dependent variable is placed on the *y*-axis (the vertical axis). When visually examining the points on the graph, if the points model a linear relationship, or a line of best fit can be drawn through the points with the points relatively close on either side, then a correlation exists. If the line of best fit has a positive slope (rises from left to right), then the variables have a positive correlation. If, like the image below, the line of best fit has a negative slope (falls from left to right), then the variables have a negative correlation. If a line of best fit cannot be drawn, then no correlation exists. A positive or negative correlation can be categorized as strong or weak, depending on how closely the points are graphed around the line of best fit.

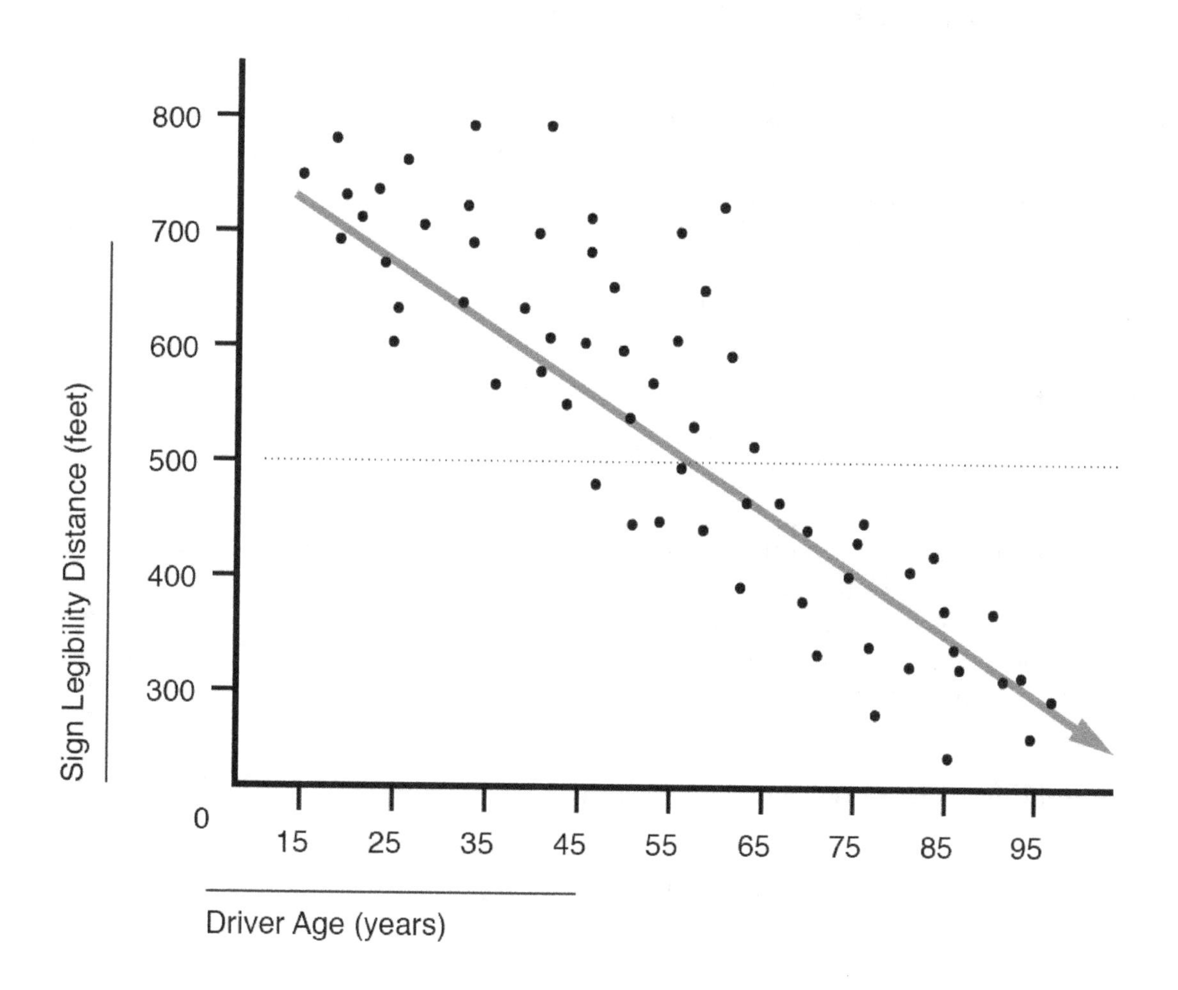

Probabilities Relative to Likelihood of Occurrence

Probability is a measure of how likely an event is to occur. Probability is written as a fraction or decimal between zero and one. If an event has a probability of zero, the event will never occur. If an event has a probability of one, the event will definitely occur. If the probability of an event is closer to zero, the event is unlikely to occur. If the probability of an event is closer to one, the event is more likely to occur. For example, a probability of $\frac{1}{2}$ means that the event is equally as likely to occur as it is not to occur. An example of this is tossing a coin. To calculate the probability of an event, the number of favorable outcomes is divided by the number of total outcomes. For example, suppose you have 2 raffle tickets out of 20 total tickets sold. The probability that you win the raffle is calculated:

$$\frac{number\ of\ favorable\ outcomes}{total\ numberof\ outcomes} = \frac{2}{20} = \frac{1}{10} \text{ (always reduce fractions)}$$

Therefore, the probability of winning the raffle is $\frac{1}{10}$ or 0.1.

Chance is the measure of how likely an event is to occur, written as a percent. If an event will never occur, the event has a 0% chance. If an event will certainly occur, the event has a 100% chance. If an event will sometimes occur, the event has a chance somewhere between 0% and 100%. To calculate chance, probability is calculated, and the fraction or decimal is converted to a percent.

The probability of multiple events occurring can be determined by multiplying the probability of each event. For example, suppose you flip a coin with heads and tails, and roll a six-sided die numbered one through six. To find the probability that you will flip heads AND roll a two, the probability of each event is determined, and those fractions are multiplied. The probability of flipping heads is $\frac{1}{2}\left(\frac{1\ side\ with\ heads}{2\ sides\ total}\right)$, and the probability of rolling a two is:

$$\frac{1}{6}\left(\frac{1\ side\ with\ a\ 2}{6\ total\ sides}\right)$$

The probability of flipping heads AND rolling a 2 is:

$$\frac{1}{2} \times \frac{1}{6} = \frac{1}{12}$$

The above scenario with flipping a coin and rolling a die is an example of independent events. **Independent events** are circumstances in which the outcome of one event does not affect the outcome of the other event. Conversely, **dependent events** are ones in which the outcome of one event affects the outcome of the second event. Consider the following scenario: a bag contains 5 black marbles and 5 white marbles. What is the probability of picking 2 black marbles without replacing the marble after the first pick?

The probability of picking a black marble on the first pick is:

$$\frac{5}{10}\left(\frac{5\ black\ marbles}{10\ total\ marbles}\right)$$

Assuming that a black marble was picked, there are now 4 black marbles and 5 white marbles for the second pick. Therefore, the probability of picking a black marble on the second pick is:

$$\frac{4}{9}\left(\frac{4\ black\ marbles}{9\ total\ marbles}\right)$$

To find the probability of picking two black marbles, the probability of each is multiplied:

$$\frac{5}{10} \times \frac{4}{9} = \frac{20}{90} = \frac{2}{9}$$

Problem Solving

Modeling and Solving Problems Using Simple Diagrams, Flowcharts, or Algorithms

Constructing, Using, and Interpreting Simple Diagrams to Solve Problems

In solving multi-step problems, the first step is to line up the available information. Then, try to decide what information the problem is asking to be found. Sketching out a diagram to illustrate what is known and what is unknown can be helpful in tackling the problem and determining what mathematical equations or processes need to be used to find the solution.

For example, you can construct a **strip diagram** to display the known information along with any information to be calculated. Finally, the missing information can be represented by a **variable** (a letter from the alphabet that represents a number) in a mathematical equation that the student can solve.

For example, Delilah collects stickers, and her friends gave her some stickers to add to her current collection. Joe gave her 45 stickers, and Aimee gave her 2 times the number of stickers that Joe gave Delilah. How many stickers did Delilah have to start with, if after her friends gave her more stickers, she had a total of 187 stickers?

In order to solve this, the given information must first be sorted out. Joe gives Delilah 45 stickers, Aimee gives Delilah 2 times the number Joe gives (2×45), and the end total of stickers is 187.

A strip diagram represents these numbers as follows:

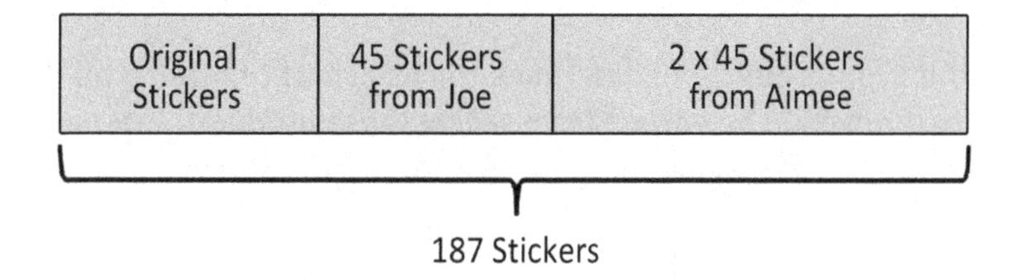

The entire situation can be modeled by this equation, using the variable s to stand for the original number of stickers:

$$s + 45 + (2 \times 45) = 187$$

Solving for s would give the solution, as follows:

$$s + 45 + 90 = 187$$

$$s + 135 = 187$$

$$s + 135 - 135 = 187 - 135$$

$$s = 52 \text{ stickers.}$$

Applying a Given Algorithm to Solve a Problem

When attempting to solve a math problem, it's important to apply the correct algorithm. It is much more difficult to determine what algorithm is necessary when solving word problems, because the necessary operations and equations are typically not provided. In these instances, the test taker must translate the words in the problem into true mathematical statements that can be solved.

Word problems can appear daunting, but don't let the wording intimidate you. No matter the scenario or specifics, the key to answering them is to translate the words into a math problem. Always keep in mind what the question is asking and what operations could lead to that answer. The following word problem resembles one of the question types most frequently encountered on the exam.

Walter's Coffee Shop sells a variety of drinks and breakfast treats.

Price List	
Hot Coffee	$2.00
Slow-Drip Iced Coffee	$3.00
Latte	$4.00
Muffin	$2.00
Crepe	$4.00
Egg Sandwich	$5.00

Costs	
Hot Coffee	$0.25
Slow-Drip Iced Coffee	$0.75
Latte	$1.00
Muffin	$1.00
Crepe	$2.00
Egg Sandwich	$3.00

Walter's utilities, rent, and labor costs him $500 per day. Today, Walter sold 200 hot coffees, 100 slow-drip iced coffees, 50 lattes, 75 muffins, 45 crepes, and 60 egg sandwiches. What was Walter's total profit today?

To accurately answer this type of question, determine the total cost of making his drinks and treats, then determine how much revenue he earned from selling those products. After arriving at these two totals, the profit is measured by deducting the total cost from the total revenue.

Walter's costs for today:

Item	Quantity	Cost Per Unit	Total Cost
Hot Coffee	200	$0.25	$50
Slow-Drip Iced Coffee	100	$0.75	$75
Latte	50	$1.00	$50
Muffin	75	$1.00	$75
Crepe	45	$2.00	$90
Egg Sandwich	60	$3.00	$180
Utilities, rent, and labor			$500
Total Costs			$1,020

Walter's revenue for today:

Item	Quantity	Revenue Per Unit	Total Revenue
Hot Coffee	200	$2.00	$400
Slow-Drip Iced Coffee	100	$3.00	$300
Latte	50	$4.00	$200
Muffin	75	$2.00	$150
Crepe	45	$4.00	$180
Egg Sandwich	60	$5.00	$300
Total Revenue			$1,530

Walter's Profit = *Revenue* – *Costs* = $1,530 – $1,020 = $510

This strategy is applicable to other question types. For example, calculating salary after deductions, balancing a checkbook, and calculating a dinner bill are common word problems similar to business planning. Just remember to use the correct operations. When a balance is increased, use addition. When a balance is decreased, use subtraction. Common sense and organization are your greatest assets when answering word problems.

In general, when solving word problems (also called story problems), it's important to understand what is being asked and to properly set up the initial equation. Always read the entire problem through, and then separate out what information is given in the statement. Decide what you are being asked to find and label each quantity with a variable or constant. Then write an equation to determine the unknown variable. Remember to label answers; sometimes knowing what the answers' units can help eliminate other possible solutions.

When trying to solve any word problem, look for a series of key words indicating addition, subtraction, multiplication, or division to help you determine how to set up the problem:

Addition: *add, altogether, together, plus, increased by, more than, in all, sum*, and *total*

Subtraction: *minus, less than, difference, decreased by, fewer than, remain*, and *take away*

Multiplication: *times, twice, of, double*, and *triple*

Division: *divided by, cut up, half, quotient of, split*, and *shared equally*

If a question asks to give words to a mathematical expression and says "equals," then an = sign must be included in the answer. Similarly, "less than or equal to" is expressed by the inequality symbol ≤, and "greater than or equal" to is expressed as ≥. Furthermore, "less than" is represented by <, and "greater than" is expressed by >.

Example:

> A store is having a spring sale, where everything is 70% off. You have $45.00 to spend. A jacket is regularly priced at $80.00. Do you have enough to buy the jacket and a pair of gloves, regularly priced at $20.00?

There are two ways to approach this.

Method 1:

Set up the equations to find the sale prices: the original price minus the amount discounted.
$\$80.00 - (\$80.00(0.70))$ = sale cost of the jacket.
$\$20.00 - (\$20.00(0.70))$ = sale cost of the gloves.
Solve for the sale cost.
$\$24.00$ = sale cost of the jacket.
$\$6.00$ = sale cost of the gloves.
Determine if you have enough money for both.
$\$24.00 + \$6.00$ = total sale cost.
$30.00 is less than $45.00, so you can afford to purchase both.

Method 2:

Determine the percent of the original price that you will pay.
$100\% - 70\% = 30\%$
Set up the equations to find the sale prices.
$\$80.00(0.30)$ = cost of the jacket.
$\$20.00(0.30)$ = cost of the gloves.
Solve.
$\$24.00$ = cost of the jacket.
$\$6.00$ = cost of the gloves.
Determine if you have enough money for both.
$\$24.00 + \$6.00$ = total sale cost.
$30.00 is less than $45.00, so you can afford to purchase both.

Example:

> Mary and Dottie team up to mow neighborhood lawns. If Mary mows 2 lawns per hour and the two of them can mow 17.5 lawns in 5 hours, how many lawns does Dottie mow per hour?

Given rate for Mary.

$$Mary = \frac{2\ lawns}{1\ hour}$$

Unknown rate of D for Dottie.

$$Dottie = \frac{D\ lawns}{1\ hour}$$

Given rate for both.

$$Total\ mowed\ together = \frac{17.5\ lawns}{5\ hours}$$

Set up the equation for what is being asked.

$$Mary + Dottie = total\ together.$$

Fill in the givens.

$$2 + D = \frac{17.5}{5}$$

Divide.

$$2 + D = 3.5$$

Subtract 2 from both sides to isolate the variable.

$$2 - 2 + D = 3.5 - 2$$

Solve and label Dottie's mowing rate.

$$D = 1.5\ lawns\ per\ hour$$

Quantitative Reasoning Practice Questions

Word Problems

1. Kassidy drove for 3 hours at a speed of 60 miles per hour. Using the distance formula, $d = r \times t$ $(distance = rate \times time)$, how far did Kassidy travel?
   a. 20 miles
   b. 180 miles
   c. 65 miles
   d. 120 miles

2. Which of the following is equivalent to the value of the digit 3 in the number 792.134?
   a. 3×10
   b. 3×100
   c. $\frac{3}{10}$
   d. $\frac{3}{100}$

3. How will the following number be written in standard form:

$$(1 \times 10^4) + (3 \times 10^3) + (7 \times 10^1) + (8 \times 10^0)$$

   a. 137
   b. 13,078
   c. 1,378
   d. 8,731

4. How will the number 847.89632 be written if rounded to the nearest hundredth?
   a. 847.90
   b. 900
   c. 847.89
   d. 847.896

5. Karen gets paid a weekly salary and a commission for every sale that she makes. The table below shows the number of sales and her pay for different weeks.

Sales	2	7	4	8
Pay	$380	$580	$460	$620

Which of the following equations represents Karen's weekly pay?
   a. $y = 90x + 200$
   b. $y = 90x - 200$
   c. $y = 40x + 300$
   d. $y = 40x - 300$

6. The phone bill is calculated each month using the equation $c = 50g + 75$. The cost of the phone bill per month is represented by c, and g represents the gigabytes of data used that month. What is the value and interpretation of the slope of this equation?

a. 75 dollars per day
b. 75 gigabytes per day
c. 50 dollars per day
d. 50 dollars per gigabyte

7. Katie works at a clothing company and sold 192 shirts over the weekend. One third of the shirts that were sold were patterned, and the rest were solid. Which mathematical expression would calculate the number of solid shirts Katie sold over the weekend?

a. $192 \times \frac{1}{3}$

b. $192 \div \frac{1}{3}$

c. $192 \times (1 - \frac{1}{3})$

d. $192 \div 3$

8. Which of the following equations best represents the problem below?

The width of a rectangle is 2 centimeters less than the length. If the perimeter of the rectangle is 44 centimeters, then what are the dimensions of the rectangle?

a. $2l + 2(l - 2) = 44$
b. $(l + 2) + (l + 2) + l = 48$
c. $l \times (l - 2) = 44$
d. $(l + 2) + (l + 2) + l = 44$

9. A company invests $50,000 in a building where they can produce saws. If the cost of producing one saw is $40, then which function expresses the amount of money the company pays? The variable y is the money paid and x is the number of saws produced.

a. $y = 50{,}000x + 40$
b. $y + 40 = x - 50{,}000$
c. $y = 40x - 50{,}000$
d. $y = 40x + 50{,}000$

10. A piggy bank contains 12 dollars' worth of nickels. A nickel weighs 5 grams, and the empty piggy bank weighs 1050 grams. What is the total weight of the full piggy bank?

a. 1,110 grams
b. 1,200 grams
c. 2,250 grams
d. 2,200 grams

11. A construction company is building a new housing development with the property of each house measuring 30 feet wide. If the length of the street is zoned off at 345 feet, how many houses can be built on the street?

a. 11
b. 115
c. 11.5
d. 12

The following stem-and-leaf plot shows plant growth in cm for a group of tomato plants.

Stem	Leaf
2	0 2 3 6 8 8 9
3	2 6 7 7
4	7 9
5	4 6 9

12. What is the range of measurements for the tomato plants' growth?

a. 29 cm
b. 37 cm
c. 39 cm
d. 59 cm

13. How many plants grew more than 35 cm?

a. 4 plants
b. 5 plants
c. 8 plants
d. 9 plants

14. At the beginning of the day, Xavier has 20 apples. At lunch, he meets his sister Emma and gives her half of his apples. After lunch, he stops by his neighbor Jim's house and gives him 6 of his apples. He then uses $\frac{3}{4}$ of his remaining apples to make an apple pie for dessert at dinner. At the end of the day, how many apples does Xavier have left?

a. 4
b. 6
c. 2
d. 1

15. It costs Shea $12 to produce 3 necklaces. If he can sell each necklace for $20, how much profit would he make if he sold 60 necklaces?
   a. $240
   b. $360
   c. $960
   d. $1200

16. Five students took a test. Jenny scored the highest with a 94. James scored the lowest with a 79. Hector scored lower than Jenny, but higher than Sam. Sam scored lower than Mary who scored an 84. Which of the following statements must be true?
   a. There were 3 people who scored higher than Sam.
   b. The median test score was an 84.
   c. Jenny is the only student who scored above 90.
   d. Hector scored lower than Mary.

17. The width of a rectangular house is 22 feet. What is the perimeter of this house if it has the same area as a house that is 33 feet wide and 50 feet long?
   a. 184 feet
   b. 200 feet
   c. 194 feet
   d. 206 feet

18. Because of an increase in demand, the price of a designer purse has increased 25% from the original price of $128. What is the new price of the purse?
   a. $32
   b. $160
   c. $192
   d. $96

19. Carly purchased 84 bulbs for her flower garden. Tulips came in trays containing six bulbs and daffodils came in trays containing 8 bulbs. Carly bought an equal number of tulip and daffodil trays. How many of each type of flower bulb were purchased?
   a. 3 trays
   b. 4 trays
   c. 6 trays
   d. 8 trays

20. Keith's bakery had 252 customers go through its doors last week. This week, that number increased to 378. Express this increase as a percentage.
   a. 26%
   b. 50%
   c. 35%
   d. 12%

21. If Danny takes 48 minutes to walk 3 miles, how long should it take him to walk 5 miles maintaining the same speed?

  a. 32 min
  b. 64 min
  c. 80 min
  d. 96 min

22. The area of a given rectangle is 24 square centimeters. If the measure of each side is multiplied by 3, what is the area of the new figure?

  a. 48 cm
  b. 72 cm
  c. 216 cm
  d. 13,824 cm

Quantitative Comparison

For each of questions 23–37, compare Quantity A to Quantity B, using additional information presented above the two quantities.

23. *h* is an integer in the following mathematical series: 4, h, 19, 39, 79

Quantity A	Quantity B
The value of *h*	9

  a. Quantity A is greater
  b. Quantity B is greater
  c. The two quantities are equal
  d. The relationship cannot be determined from the information given.

24.

g inches

Area = 56 square inches

4 inches

Quantity A
The value of g

Quantity B
13

a. Quantity A is greater
b. Quantity B is greater
c. The two quantities are equal
d. The relationship cannot be determined from the information given.

25. $4x - 12 = -2x$

Quantity A
The value of x

Quantity B
3

a. Quantity A is greater
b. Quantity B is greater
c. The two quantities are equal
d. The relationship cannot be determined from the information given.

26.

Jimmy	Steve
7 red marbles	6 green marbles
8 blue marbles	4 blue marbles

Quantity A
All of Jimmy's marbles divided by all of Steve's marbles

Quantity B
Jimmy's blue marbles divided by Steve's green marbles

a. Quantity A is greater
b. Quantity B is greater
c. The two quantities are equal
d. The relationship cannot be determined from the information given.

27.

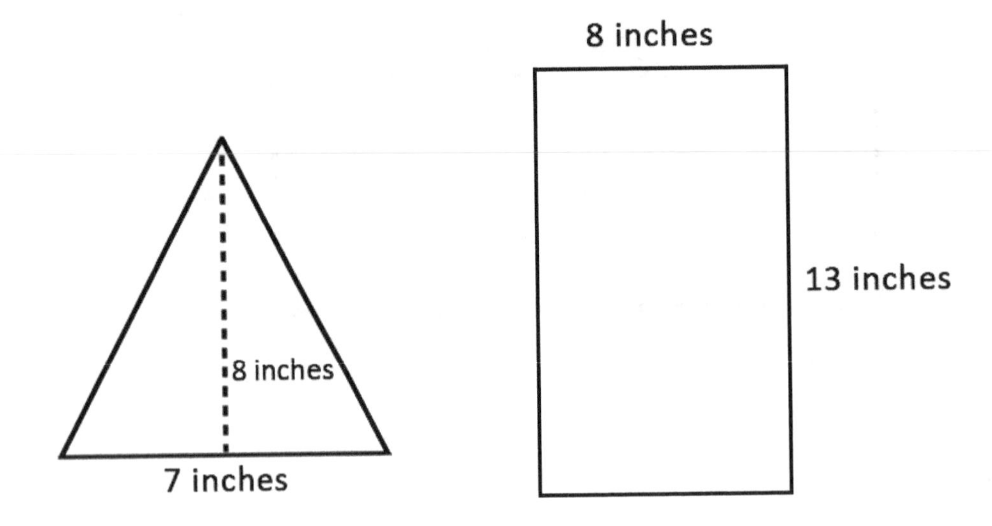

Quantity A	Quantity B
7 times the area of the triangle	2 times the area of the rectangle

a. Quantity A is greater
b. Quantity B is greater
c. The two quantities are equal
d. The relationship cannot be determined from the information given.

28. Truck A drives 1236 yards and truck B drives 3680 feet.

Quantity A	Quantity B
The distance that truck *A* drove	The distance that truck *B* drove

a. Quantity A is greater
b. Quantity B is greater
c. The two quantities are equal
d. The relationship cannot be determined from the information given.

29. $x > 6 > z$

Quantity A	Quantity B
$x + z$	$x - 6$

a. Quantity A is greater
b. Quantity B is greater
c. The two quantities are equal
d. The relationship cannot be determined from the information given.

30. There are 16 rocks in a bag. 12 of them are smooth and 4 of them are rough.

Quantity A	Quantity B
The probability of choosing a rough rock	$\frac{2}{8}$

a. Quantity A is greater
b. Quantity B is greater
c. The two quantities are equal
d. The relationship cannot be determined from the information given.

31. Bill is four years older than Jim.

Quantity A	Quantity B
Twice Jim's age	Bill's age

a. Quantity A is greater
b. Quantity B is greater
c. The two quantities are equal
d. The relationship cannot be determined from the information given.

32. Angie has more cats than Janet.

Quantity A	Quantity B
Angie's number of cats	4 more than Janet's number of cats

a. Quantity A is greater
b. Quantity B is greater
c. The two quantities are equal
d. The relationship cannot be determined from the information given.

33. Gage is twice as old as Cam.

Quantity A	Quantity B
Cam's age	Half of Gage's age

a. Quantity A is greater
b. Quantity B is greater
c. The two quantities are equal
d. The relationship cannot be determined from the information given.

34.

Quantity A	Quantity B
Largest prime number less than 35	Smallest prime number greater than 25

a. Quantity A is greater
b. Quantity B is greater
c. The two quantities are equal
d. The relationship cannot be determined from the information given.

35.

Quantity A	Quantity B
28% of 345	$\frac{1}{5}$ of 300

a. Quantity A is greater
b. Quantity B is greater
c. The two quantities are equal
d. The relationship cannot be determined from the information given.

36.

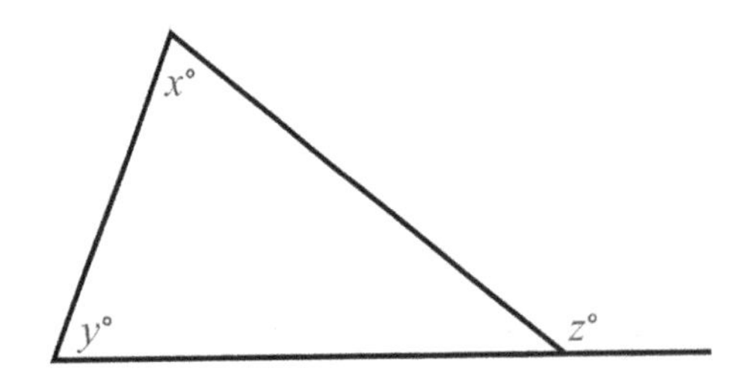

Quantity A	Quantity B
$x + y$	z

a. Quantity A is greater
b. Quantity B is greater
c. The two quantities are equal
d. The relationship cannot be determined from the information given.

37.

Quantity A	Quantity B
Circumference of a circle with radius 4 cm	Perimeter of a rectangle with sides 5 cm and 7 cm

a. Quantity A is greater
b. Quantity B is greater
c. The two quantities are equal
d. The relationship cannot be determined from the information given.

Quantitative Reasoning Answer Explanations

Word Problems

1. B: 180 miles. The rate, 60 miles per hour, and time, 3 hours, are given for the scenario. To determine the distance traveled, the given values for the rate (*r*) and time (*t*) are substituted into the distance formula and evaluated:

$$d = r \times t$$

$$d = \left(\frac{60mi}{h}\right) \times (3h)$$

$$d = 180mi$$

2. D: $\frac{3}{100}$. Each digit to the left of the decimal point represents a higher multiple of 10 and each digit to the right of the decimal point represents a quotient of a higher multiple of 10 for the divisor. The first digit to the right of the decimal point is equal to the value $\div$ 10. The second digit to the right of the decimal point is equal to the value $\div$ (10×10), or the value $\div$ 100.

3. B: 13,078. The power of 10 by which a digit is multiplied corresponds with the number of zeros following the digit when expressing its value in standard form. Therefore:

$$(1 \times 10^4) + (3 \times 10^3) + (7 \times 10^1) + (8 \times 10^0)$$

$$10{,}000 + 3{,}000 + 70 + 8 = 13{,}078$$

4. A: 847.90. The hundredths place value is located two digits to the right of the decimal point (the digit 9 in the original number). The digit to the right of the place value is examined to decide whether to round up or keep the digit. In this case, the digit 6 is 5 or greater so the hundredth place is rounded up. When rounding up, if the digit to be increased is a 9, the digit to its left is increased by one and the digit in the desired place value is made a zero. Therefore, the number is rounded to 847.90.

5. C: $y = 40x + 300$

In this scenario, the variables are the number of sales and Karen's weekly pay. The weekly pay depends on the number of sales. Therefore, weekly pay is the dependent variable (*y*) and the number of sales is the independent variable (*x*). Each pair of values from the table can be written as an ordered pair (*x*, *y*): (2, 380), (7, 580), (4, 460), (8, 620). The ordered pairs can be substituted into the equations to see which creates true statements (both sides equal) for each pair. Even if one ordered pair produces equal values for a given equation, the other three ordered pairs must be checked. The only equation which is true for all four ordered pairs is $y = 40x + 300$:

$$380 = 40(2) + 300 \rightarrow 380 = 380$$

$$580 = 40(7) + 300 \rightarrow 580 = 580$$

$$460 = 40(4) + 300 \rightarrow 460 = 460$$

$$620 = 40(8) + 300 \rightarrow 620 = 620$$

6. D: The slope from this equation is 50, and it is interpreted as the cost per gigabyte used. Since the *g*-value represents number of gigabytes and the equation is set equal to the cost in dollars, the slope relates these two values. For every gigabyte used on the phone, the bill goes up 50 dollars.

7. C: $\frac{1}{3}$ of the shirts sold were patterned. Therefore, $1 - \frac{1}{3} = \frac{2}{3}$ of the shirts sold were solid. Anytime "of" a quantity appears in a word problem, multiplication should be used. Therefore:

$$192 \times \frac{2}{3} = \frac{192 \times 2}{3} = \frac{384}{3} = 128 \text{ solid shirts were sold}$$

The entire expression is $192 \times \left(1 - \frac{1}{3}\right)$.

8. A: The first step is to determine the unknown, which is in terms of the length, l.

The second step is to translate the problem into the equation using the perimeter of a rectangle:

$$P = 2l + 2w$$

The width is the length minus 2 centimeters. The resulting equation is:

$$2l + 2(l - 2) = 44$$

The equation can be solved as follows:

$2l + 2l - 4 = 44$	Apply the distributive property on the left side of the equation
$4l - 4 = 44$	Combine like terms on the left side of the equation
$4l = 48$	Add 4 to both sides of the equation
$l = 12$	Divide both sides of the equation by 4

The length of the rectangle is 12 centimeters. The width is the length minus 2 centimeters, which is 10 centimeters. Checking the answers for length and width forms the following equation:

$$44 = 2(12) + 2(10)$$

The equation can be solved using the order of operations to form a true statement: $44 = 44$.

9. D: For manufacturing costs, there is a linear relationship between the cost to the company and the number produced, with a *y*-intercept given by the base cost of acquiring the means of production, and a slope given by the cost to produce one unit. In this case, that base cost is $50,000, while the cost per unit is $40. So:

$$y = 40x + 50{,}000$$

10. C: A dollar contains 20 nickels. Therefore, if there are 12 dollars' worth of nickels, there are:

$$12 \times 20 = 240 \text{ nickels}$$

Each nickel weighs 5 grams. Therefore, the weight of the nickels is:

$$240 \times 5 = 1{,}200 \text{ grams}$$

Adding in the weight of the empty piggy bank, the filled bank weighs 2,250 grams.

11. A: 11. To determine the number of houses that can fit on the street, the length of the street is divided by the width of each house:

$$345 \div 30 = 11.5$$

Although the mathematical calculation of 11.5 is correct, this answer is not reasonable. Half of a house cannot be built, so the company will need to either build 11 or 12 houses. Since the width of 12 houses (360 feet) will extend past the length of the street, only 11 houses can be built.

12. C: The range of the entire stem-and-leaf plot is found by subtracting the lowest value from the highest value, as follows:

$$59 - 20 = 39 \text{ cm}$$

All other choices are miscalculations read from the chart.

13. C: To calculate the total greater than 35, the number of measurements above 35 must be totaled; 36, 37, 37, 47, 49, 54, 56, 59 = 8 measurements. Choice *A* is the number of measurements in the 3 categories, Choice *B* is the number in the 4 and 5 categories, and Choice *D* is the number in the 3, 4, and 5 categories.

14. D: This problem can be solved using basic arithmetic. Xavier starts with 20 apples, then gives his sister half, so 20 divided by 2.

$$\frac{20}{2} = 10$$

He then gives his neighbor 6, so 6 is subtracted from 10.

$$10 - 6 = 4$$

Lastly, he uses ¾ of his apples to make an apple pie, so to find remaining apples, the first step is to subtract ¾ from one and then multiply the difference by 4.

$$\left(1 - \frac{3}{4}\right) \times 4 = ?$$

$$\left(\frac{4}{4} - \frac{3}{4}\right) \times 4 = ?$$

$$\left(\frac{1}{4}\right) \times 4 = 1$$

15. C: In order to calculate the profit, we need to create an equation that models the total income minus the cost of the materials.

$$\$60 \times 20 = \$1{,}200 \text{ total income}$$

$60 \div 3 = 20$ sets of materials. $20 \times \$12 = \$240$ cost of materials.

$$\$1{,}200 - \$240 = \$960 \text{ profit}$$

Choice *A* is not correct, as it is only the cost of materials. Choice *B* is not correct, as it is a miscalculation. Choice *D* is not correct, as it is the total income from the sale of the necklaces.

16. A: It can be determined from reading the information given that Jenny, Hector, and Mary scored higher than Sam, so Choice *A* is correct. There is no relation provided between Hector and Mary's scores. Given that Mary could have scored higher or lower than Hector, it cannot be determined if her score is the median, so Choice *B* is incorrect. With the information given, it is possible that Hector scored above 90, so Choice *C* is incorrect. There is no relation given between Hector and Mary's scores. This means that Hector could have scored higher or lower than Mary. So, Choice *D* is incorrect.

17. C: First, find the area of the second house. The area is:

$$A = l\,x\,w = 33 \times 50 = 1{,}650 \text{ square feet}$$

Then use the area formula to determine what length gives the first house an area of 1,650 square feet. So:

$$1{,}650 = 22 \times l, l = \frac{1{,}650}{22} = 75 \text{ feet}$$

Then, use the formula for perimeter to get:

$$75 + 75 + 22 + 22 = 194 \text{ feet}$$

18. B: The new price of the purse can be found by first multiplying the original price by 25%, or 0.25. This yields an increase of $32. Taking the original price of $128 and adding the increase in price of $32 yields a new price of $160.

19. C: This problem involved setting up an algebraic equation to solve for x, or the number of flower trays Carly purchased. The equation is as follows:

$$6x + 8x = 84$$

So,

$$14x = 84$$

Then divide each side by 14 to solve for x:

$$x = \frac{84}{14} = 6\ trays$$

20. B: First, calculate the difference between the larger value and the smaller value.

$$378 - 252 = 126$$

To calculate this difference as a percentage of the original value, and thus calculate the percentage increase, divide 126 by 252, then multiply by 100 to reach the percentage = 50%, answer *B*.

21. C: 80 min. To solve the problem, a proportion is written consisting of ratios comparing distance and time. One way to set up the proportion is:

$$\frac{3}{48} = \frac{5}{x}\left(\frac{distance}{time} = \frac{distance}{time}\right)$$

where x represents the unknown value of time. To solve a proportion, the ratios are cross-multiplied:

$$(3)(x) = (5)(48) \rightarrow 3x = 240$$

The equation is solved by isolating the variable, or dividing by 3 on both sides, to produce $x = 80$.

22. C: 216 cm. Because area is a two-dimensional measurement, the dimensions are multiplied by a scale that is squared to determine the scale of the corresponding areas. The dimensions of the rectangle are multiplied by a scale of 3. Therefore, the area is multiplied by a scale of 3^2 (which is equal to 9):

$$24cm \times 9 = 216cm$$

Quantitative Comparison

23. C: The equation that produces this series is $2x + 1$. This gives $2(4) + 1 = 9, 2(9) + 1 = 19$, and so on. This means that the value of h in the series is 9, so Quantity A and Quantity B are equal.

24. A: The value of g can be found using the formula for area of a rectangle ($A = l \times w$). So, $56 = g \times 4$, and $g = 14$. This means that Quantity A is greater than Quantity B.

25. B: The first step is to solve for x. For this equation that is:

$$4x - 12 = -2x$$

$$6x - 12 = 0$$

$$6x = 12$$

$$x = 2$$

Since the value of x is 2 and Quantity B is 3, it means that Quantity B is greater.

26. A: The first step here is to solve each of the ratios. The first ratio is all of Jimmy's marbles divided by all of Steve's marbles. This gives:

$$\frac{15}{10} = \frac{3}{2}$$

The second ratio is all of Jimmy's blue marbles divided by all of Steve's green marbles. This gives:

$$\frac{8}{6} = \frac{4}{3}$$

Since $\frac{3}{2}$ is greater than $\frac{4}{3}$, Quantity A is greater.

27. B: First, find the area of both figures. The area of the triangle is:

$$\frac{1}{2}(7) \times 8 = 28 \text{ square inches}$$

The area of the rectangle is $13 \times 8 = 104$ square inches. So, 7 times the area of the triangle would be 196 square inches, and 2 times the area of the rectangle would be 208 square inches. This means that Quantity B is greater.

28. A: First, convert the distance that Truck A drove to feet. This is:

$$1{,}236 \times 3 = 3{,}708 \text{ feet}$$

This means that Truck A drove further than Truck B. So, Quantity A is greater than Quantity B.

29. D: There is not enough information to determine which quantity is greater. Either quantity could be greater for given values of x and z. For example, if $x = 7$ and $z = 1$, then Quantity A is 8 and Quantity B is 1. If, $x = 10$ and $z = -8$, then Quantity A is 2 and Quantity B is 4.

30. C: The probability of choosing a rough rock is $\frac{4}{16}$. This is equal to $\frac{2}{8}$.

31. D: If Jim is 2 years old, then Bill is 6 years old. Twice Jim's age is 4, so Quantity B is greater. However, if Jim is 20 years old, then Bill is 24. Twice Jim's age is 40, so Quantity A is greater. Thus, the relationship cannot be determined from the given information.

32. D: If Angie has 5 cats, then Janet may have 4 cats. Four more than Janet is 8 cats, so Quantity B would be greater. If Angie has 10 cats and Janet has 2 cats, then four more cats than Janet is 6, and Quantity A is greater. Thus, there is not enough information to determine which quantity is greater.

33. C: If Cam is 10 years old, then Gage is 20 years old. Cam's age is 10, and half of Gage's age is 10, so the quantities are equal. If Gage is 50 years old, then Cam is 25. Cam's age is 25, and half of Gage's age is 25, so the quantities are equal. No matter what Cam's age is, it will always be equal to half of Gage's age. The correct answer is Choice *C*, the two quantities are equal.

34. A: A prime number is a number that only has factors of 1 and itself. Quantity A is the largest prime number less than 35. The number 34 is even, or divisible by 2, so it is not prime. The number 33 is divisible by 3 and 11, and the number 32 is even, so neither of those is prime. The number 31 has factors of only 1 and itself, so it is the largest prime number less than 35. Quantity B is the smallest prime number greater than 25. The next number is 26, which is even. The number 27 has factors of 3 and 9, while the number 28 is even. The number 29 has factors of only 1 and itself; therefore, it is the smallest prime number greater than 25. Comparing these two numbers, 31 and 29, Quantity A is greater. The correct answer is Choice *A* because Quantity A is greater.

35. A: The percent of a number can be found by using the equation $\frac{\%}{100} = \frac{is}{of}$. After filling in the values for Quantity A, this equation becomes $\frac{28}{100} = \frac{x}{345}$. Since the "is," or x-portion, of the equation is missing, it can be solved using cross multiplication. This yields the equation:

$$28 \times 345 = 100 \times x$$

Multiplying 28 by 345 and then dividing by 100 gives an x-value of 96.6. Quantity B is $\frac{1}{5}$ of 300. Splitting 300 into 5 equal parts of 60 means Quantity B is equal to 60. The value of 96.6 is greater than 60. Therefore, Quantity A is greater than Quantity B. The correct answer is *A*. Quantity A, with a value of 96.6, is greater than Quantity B, with a value of 60.

36. C: The angles inside a triangle are supplementary, which means they add up to 180 degrees. The third angle not named in the triangle and the exterior angle *z* are also supplementary because they add up to a straight line, or 180 degrees. If the unnamed angle is given the value *w*, then the equations can be written: $x + y + w = 180$ and $w + z = 180$. By the transitive property of equality, the equation can be written:

$$x + y + w = w + z$$

Subtracting w from both sides yields the equation $x + y = z$. Using this equation, Quantity A and Quantity B can be stated as equal. The correct answer is *C* because Quantity B is equal to Quantity A by the proof given above.

37. A: The circumference of a circle can be found by using the formula $C = 2\pi r$. Using this formula with the radius of 4 cm produces the circumference:

$$C = 2 \times \pi \times 4 = 25.12\ cm$$

The perimeter of a rectangle can be found by using the formula:

$$P = 2l + 2w$$

Plugging in the values for the sides given produces the equation:

$$P = 2(5) + 2(7) = 24\ cm$$

The circumference in Quantity A of 25.12 cm is greater than the perimeter in Quantity B of 24 cm. These two quantities can be compared because they both describe the distance around the outside of a 2-dimensional shape. The correct answer is *A* because Quantity A is greater than Quantity B.

Mathematics Achievement Practice Questions

1. What is the value of the sum of $\frac{1}{3}$ and $\frac{2}{5}$?
   a. $\frac{3}{8}$
   b. $\frac{11}{15}$
   c. $\frac{11}{30}$
   d. $\frac{4}{5}$

2. What is the value of the expression: $7^2 - 3 \times (4 + 2) + 15 \div 5$?
   a. 12.2
   b. 40.2
   c. 34
   d. 58.2

3. How will $\frac{4}{5}$ be written as a percent?
   a. 40%
   b. 125%
   c. 90%
   d. 80%

4. If $-3(x + 4) \geq x + 8$, what is the value of x?
   a. $x = 4$
   b. $x \geq 2$
   c. $x \geq -5$
   d. $x \leq -5$

5. What is the 42nd item in the pattern: ▲oo□▲oo□▲...?
   a. o
   b. ▲
   c. □
   d. None of the above

6. A closet is filled with red, blue, and green shirts. If $\frac{1}{3}$ of the shirts are green and $\frac{2}{5}$ are red, what fraction of the shirts are blue?
   a. $\frac{4}{15}$
   b. $\frac{1}{5}$
   c. $\frac{7}{15}$
   d. $\frac{1}{2}$

7. Shawna buys $2\frac{1}{2}$ gallons of paint. If she uses $\frac{1}{3}$ of it on the first day, how much does she have left?

a. $1\frac{5}{6}$ gallons

b. $1\frac{1}{2}$ gallons

c. $1\frac{2}{3}$ gallons

d. 2 gallons

8. Solve this equation:

$$9x + x - 7 = 16 + 2x$$

a. $x = -4$

b. $x = 3$

c. $x = \frac{9}{8}$

d. $x = \frac{23}{8}$

9. Arrange the following numbers from least to greatest value:

$0.85, \frac{4}{5}, \frac{2}{3}, \frac{91}{100}$

a. $0.85, \frac{4}{5}, \frac{2}{3}, \frac{91}{100}$

b. $\frac{4}{5}, 0.85, \frac{91}{100}, \frac{2}{3}$

c. $\frac{2}{3}, \frac{4}{5}, 0.85, \frac{91}{100}$

d. $0.85, \frac{91}{100}, \frac{4}{5}, \frac{2}{3}$

10. Which graph will be a line parallel to the graph of $y = 3x - 2$?

a. $2y - 6x = 2$
b. $y - 4x = 4$
c. $3y = x - 2$
d. $2x - 2y = 2$

11. An equation for the line passing through the origin and the point $(2, 1)$ is

a. $y = 2x$
b. $y = \frac{1}{2}x$
c. $y = x - 2$
d. $2y = x + 1$

12. A line goes through the point (-4, 0) and the point (0, 2). What is the slope of the line?

a. 2

b. 4

c. $\frac{3}{2}$

d. $\frac{1}{2}$

13. Write the expression for three times the sum of twice a number and one minus 6.

a. $2x + 1 - 6$
b. $3x + 1 - 6$
c. $3(x + 1) - 6$
d. $3(2x + 1) - 6$

14. What would be the coordinates if the point plotted on the grid was reflected over the x-axis?

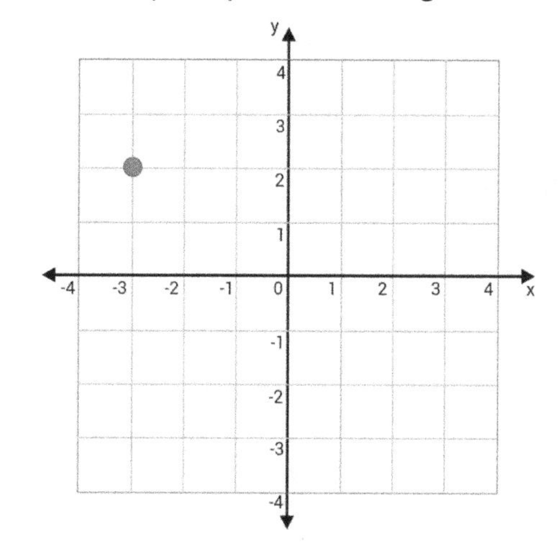

a. (-3, 2)
b. (2, -3)
c. (-3, -2)
d. (2, 3)

15. Jessica buys 10 cans of paint. Red paint costs $1 per can and blue paint costs $2 per can. In total, she spends $16. How many red cans did she buy?

a. 2
b. 3
c. 4
d. 5

16. The variable y is directly proportional to x. If $y = 3$ when $x = 5$, then what is y when $x = 20$?

a. 10
b. 12
c. 14
d. 16

17. Apples cost \$2 each, while bananas cost \$3 each. Maria purchased 10 fruits in total and spent \$22. How many apples did she buy?
   a. 5
   b. 6
   c. 7
   d. 8

18. If $-3(x+4) \geq x+8$, what is the value of x?
   a. $x = 4$
   b. $x \geq 2$
   c. $x \geq -5$
   d. $x \leq -5$

19. Dwayne has received the following scores on his math tests: 78, 92, 83, 97. What score must Dwayne get on his next math test to have an overall average of at least 90?
   a. 89
   b. 98
   c. 95
   d. 100

20. What is the overall median of Dwayne's current scores: 78, 92, 83, 97?
   a. 19
   b. 85
   c. 83
   d. 87.5

21. Simplify the following fraction:

$$\frac{\frac{5}{7}}{\frac{9}{11}}$$

   a. $\frac{55}{63}$
   b. $\frac{7}{1000}$
   c. $\frac{13}{15}$
   d. $\frac{5}{11}$

22. What is the slope of this line?

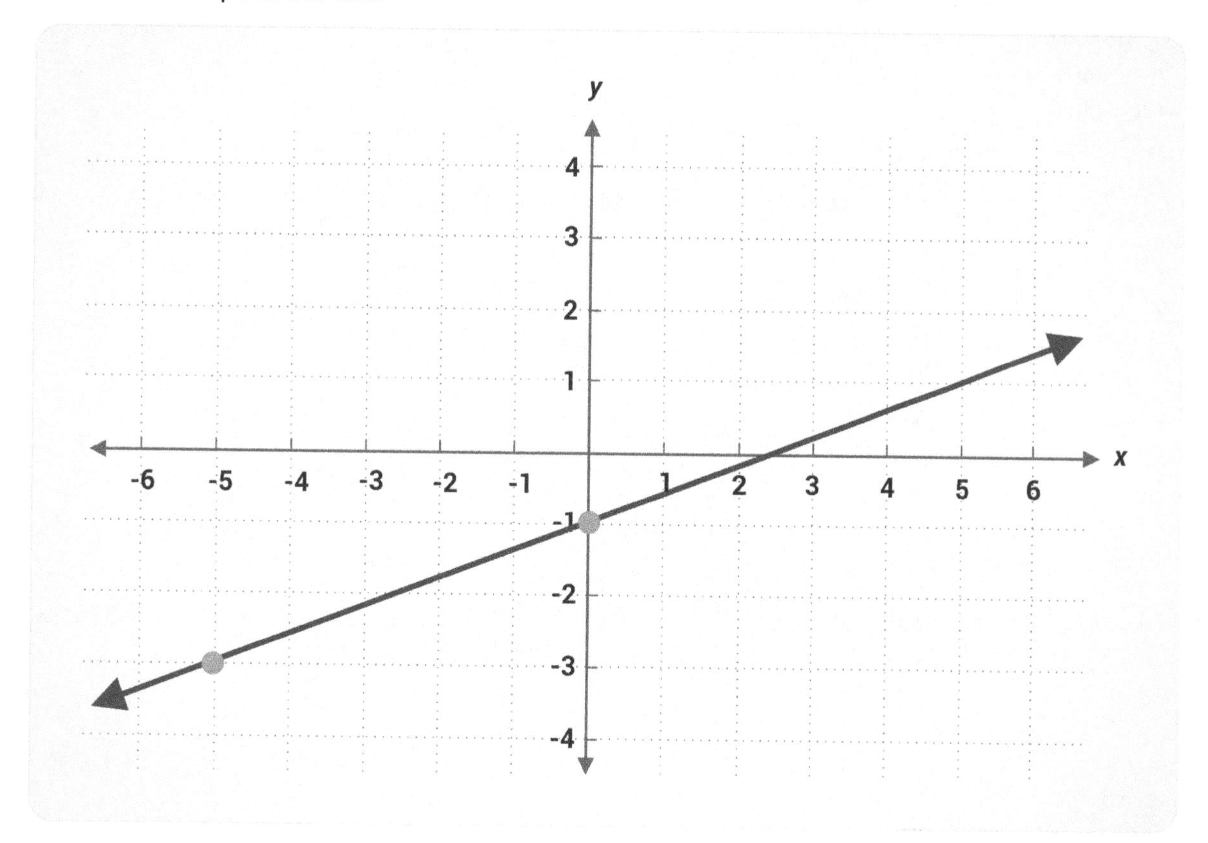

a. 2

b. $\frac{5}{2}$

c. $\frac{1}{2}$

d. $\frac{2}{5}$

23. What is the perimeter of the figure below? Note that the solid outer line is the perimeter.

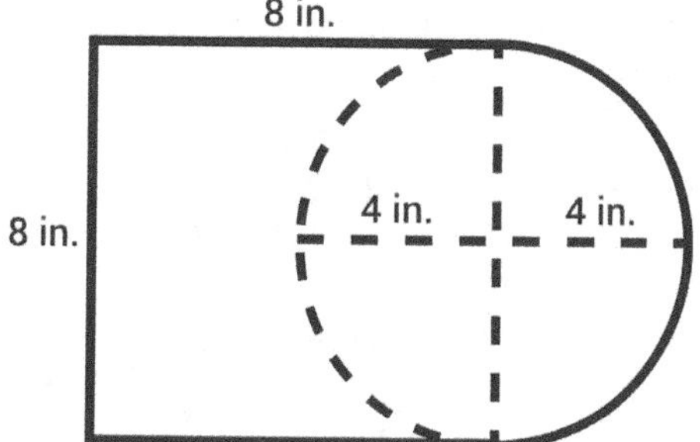

a. 48.565 in
b. 36.565 in
c. 39.78 in
d. 39.565 in

24. Which four-sided shape is always a rectangle?
 a. Rhombus
 b. Square
 c. Parallelogram
 d. Quadrilateral

25. Of the numbers -1, 17, -13, or -6, which is farthest from 5 on the number line?
 a. -1
 b. 17
 c. -13
 d. -6

26. If $\sqrt{1+x} = 4$, what is *x*?
 a. 10
 b. 15
 c. 20
 d. 25

27. Five students take a test. The scores of the first four students are 80, 85, 75, and 60. If the median score is 80, which of the following could NOT be the score of the fifth student?
 a. 60
 b. 80
 c. 85
 d. 100

28. Which of the following statements is true about the two lines below?

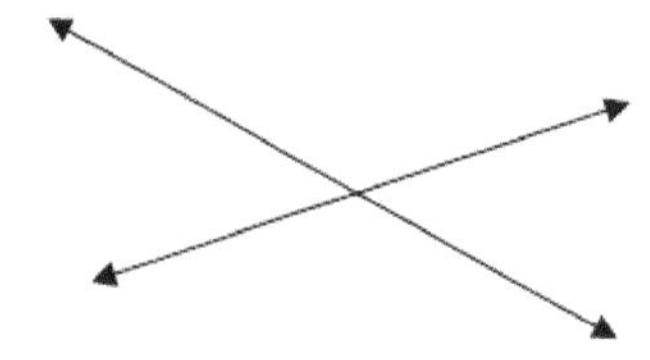

 a. The two lines are parallel but not perpendicular.
 b. The two lines are perpendicular but not parallel.
 c. The two lines are both parallel and perpendicular.
 d. The two lines are neither parallel nor perpendicular.

29. What is the volume of a rectangular prism with a height of 2 inches, a width of 4 inches, and a depth of 6 inches?
 a. 12 in^3
 b. 24 in^3
 c. 48 in^3
 d. 96 in^3

30. What is the volume of a cylinder, in terms of π, with a radius of 5 inches and a height of 10 inches?

a. $250\,\pi$ in^3
b. $50\,\pi$ in^3
c. $100\,\pi$ in^3
d. $200\,\pi$ in^3

31. Ten students take a test. Five students get a 50. Four students get a 70. If the average score is 55, what was the last student's score?

a. 20
b. 40
c. 50
d. 60

32. An equilateral triangle has a perimeter of 18 feet. If a square whose sides have the same length as one side of the triangle is built, what will be the area of the square?

a. 6 square feet
b. 36 square feet
c. 256 square feet
d. 1000 square feet

33. Which of the following numbers has the greatest value?

a. 1.4378
b. 1.07548
c. 1.43592
d. 0.89409

34. What is the solution to $(2 \times 20) \div (7 + 1) + (6 \times 0.01) + (4 \times 0.001)$?

a. 5.064
b. 5.64
c. 5.0064
d. 48.064

35. Last year, the New York City area received approximately $27\frac{3}{4}$ inches of snow. The Denver area received approximately 3 times as much snow as New York City. How much snow fell in Denver?

a. 60 inches
b. $27\frac{1}{4}$ inches
c. $9\frac{1}{4}$ inches
d. $83\frac{1}{4}$ inches

36. Solve for x: $\frac{2x}{5} - 1 = 59$.

a. 60
b. 145
c. 150
d. 115

37. A National Hockey League store in the state of Michigan advertises 50% off all items. Sales tax in Michigan is 6%. How much would a hat originally priced at $32.99 and a jersey originally priced at $64.99 cost during this sale? Round to the nearest penny.

a. $97.98
b. $103.86
c. $51.93
d. $48.99

38. Store brand coffee beans cost $1.23 per pound. A local coffee bean roaster charges $1.98 per 1 ½ pounds. How much more would 5 pounds from the local roaster cost than 5 pounds of the store brand?

a. $0.55
b. $1.55
c. $1.45
d. $0.45

39. Paint Inc. charges $2000 for painting the first 1,800 feet of trim on a house and $1.00 per foot for each foot after. How much would it cost to paint a house with 3125 feet of trim?

a. $3125
b. $2000
c. $5125
d. $3325

40. A train traveling 50 miles per hour takes a trip lasting 3 hours. If a map has a scale of 1 inch per 10 miles, how many inches apart are the train's starting point and ending point on the map?

a. 14
b. 12
c. 13
d. 15

41. A traveler takes an hour to drive to a museum, spends 3 hours and 30 minutes there, and takes half an hour to drive home. What percentage of his or her time was spent driving?

a. 15%
b. 30%
c. 40%
d. 60%

42. A couple buys a house for $150,000. They sell it for $165,000. By what percentage did the house's value increase?

a. 10%
b. 13%
c. 15%
d. 17%

43. A map has a scale of 1 inch per 5 miles. A car can travel 60 miles per hour. If the distance from the start to the destination is 3 inches on the map, how long will it take the car to make the trip?

a. 12 minutes
b. 15 minutes
c. 17 minutes
d. 20 minutes

44. Which of the following figures is not a polygon?
   a. Decagon
   b. Cone
   c. Triangle
   d. Rhombus

45. What is the area of the regular hexagon shown below?

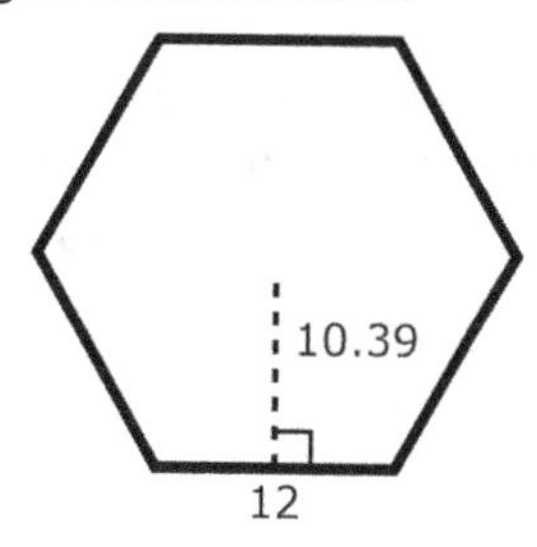

   a. 72
   b. 124.68
   c. 374.04
   d. 748.08

46. Given the value of a given stock at monthly intervals, which graph should be used to best represent the trend of the stock?
   a. Box plot
   b. Line plot
   c. Line graph
   d. Circle graph

47. What is the probability of randomly picking the winner and runner-up from a race of 4 horses and distinguishing which is the winner?
   a. $\frac{1}{4}$
   b. $\frac{1}{2}$
   c. $\frac{1}{16}$
   d. $\frac{1}{12}$

Mathematics Achievement Answer Explanations

1. B: $\frac{11}{15}$. Fractions must have like denominators to be added. We are trying to add a fraction with a denominator of 3 to a fraction with a denominator of 5, so we have to convert both fractions to their respective equivalent fractions that have a common denominator. The common denominator is the least common multiple (LCM) of the two original denominators. In this case, the LCM is 15, so both fractions should be changed to equivalent fractions with a denominator of 15. To determine the numerator of the new fraction, the old numerator is multiplied by the same number by which the old denominator is multiplied to obtain the new denominator. For the fraction $\frac{2}{5}$, multiplying both the numerator and denominator by 3 produces $\frac{6}{15}$. When fractions have like denominators, they are added by adding the numerators and keeping the denominator the same:

$$\frac{5}{15}+\frac{6}{15}=\frac{11}{15}$$

2. C: 34. When performing calculations consisting of more than one operation, the order of operations should be followed: *P*arenthesis, *E*xponents, *M*ultiplication/*D*ivision, *A*ddition/*S*ubtraction. Parenthesis:

$$7^2-3\times(4+2)+15\div 5$$

$$7^2-3\times(6)+15\div 5$$

Exponents:

$$7^2-3\times 6+15\div 5$$

$$49-3\times 6+15\div 5$$

Multiplication/Division (from left to right):

$$49-3\times 6+15\div 5$$

$$49-18+3$$

Addition/Subtraction (from left to right):

$$49-18+3=34$$

3. D: 80%. To convert a fraction to a percent, the fraction is first converted to a decimal. To do so, the numerator is divided by the denominator: $4\div 5=0.8$. To convert a decimal to a percent, the number is multiplied by 100:

$$0.8\times 100=80\%$$

4. D: $x\le -5$. When solving a linear equation or inequality:

Distribution is performed if necessary:

$$-3(x+4)\rightarrow -3x-12\ge x+8$$

This means that any like terms on the same side of the equation/inequality are combined.

The equation/inequality is manipulated to get the variable on one side. In this case, subtracting x from both sides produces:

$$-4x - 12 \geq 8$$

The variable is isolated using inverse operations to undo addition/subtraction. Adding 12 to both sides produces:

$$-4x \geq 20$$

The variable is isolated using inverse operations to undo multiplication/division. Remember if dividing by a negative number, the relationship of the inequality reverses, so the sign is flipped. In this case, dividing by -4 on both sides produces $x \leq -5$.

5. A: ○. The core of the pattern consists of 4 items: ▲○○□. Therefore, the core repeats in multiples of 4, with the pattern starting over on the next step. The closest multiple of 4 to 42 is 40. Step 40 is the end of the core (□), so step 41 will start the core over (▲) and step 42 is ○.

6. A: The total fraction taken up by green and red shirts will be:

$$\frac{1}{3} + \frac{2}{5} = \frac{5}{15} + \frac{6}{15} = \frac{11}{15}$$

The remaining fraction is:

$$1 - \frac{11}{15} = \frac{15}{15} - \frac{11}{15} = \frac{4}{15}$$

7. C: If she has used $\frac{1}{3}$ of the paint, she has $\frac{2}{3}$ remaining. $2\frac{1}{2}$ gallons are the same as $\frac{5}{2}$ gallons. The calculation is:

$$\frac{2}{3} \times \frac{5}{2} = \frac{5}{3} = 1\frac{2}{3} \text{ gallons}$$

8. D:

$9x + x - 7 = 16 + 2x$	Combine $9x$ and x.
$10x - 7 = 16 + 2x$	
$10x - 7 + 7 = 16 + 2x + 7$	Add 7 to both sides to remove (-7).
$10x = 23 + 2x$	
$10x - 2x = 23 + 2x - 2x$	Subtract 2x from both sides to move it to the other side of the equation.
$8x = 23$	
$\frac{8x}{8} = \frac{23}{8}$	Divide by 8 to get x by itself.
$x = \frac{23}{8}$	

9. C: The first step is to depict each number using decimals. $\frac{91}{100}$ = 0.91

Dividing the numerator by denominator of $\frac{4}{5}$ to convert it to a decimal yields 0.80, while $\frac{2}{3}$ becomes 0.66 recurring. Rearrange each expression in ascending order, as found in Choice *C*.

10. A: Parallel lines have the same slope. The slope of *C* can be seen to be $\frac{1}{3}$ by dividing both sides by 3. The other answers are in standard form $Ax + By = C$, for which the slope is given by $\frac{-A}{B}$. The slope of *A* is 3, the slope of *B* is 4. The slope of *D* is 1.

11. B: The slope will be given by:

$$\frac{1-0}{2-0} = \frac{1}{2}$$

The *y*-intercept will be 0, since it passes through the origin. Using slope-intercept form, the equation for this line is $y = \frac{1}{2}x$.

12. D: The slope is given by the change in *y* divided by the change in *x*. The change in *y* is $2 - 0 = 2$, and the change in *x* is $0 - (-4) = 4$. The slope is $\frac{2}{4} = \frac{1}{2}$.

13. D: The expression is three times the sum of twice a number and 1, which is $3(2x + 1)$. Then, 6 is subtracted from this expression.

14. C: (-3, -2). The coordinates of a point are written as an ordered pair (*x*, *y*). To determine the *x*-coordinate, a line is traced directly above or below the point until reaching the *x*-axis. This step notes the value on the *x*-axis. In this case, the *x*-coordinate is -3. To determine the *y*-coordinate, a line is traced directly to the right or left of the point until reaching the y-axis, which notes the value on the *y*-axis. In this case, the *y*-coordinate is 2. Therefore, the current ordered pair is written (-3, 2). To reflect the point over the *x*-axis, the *x*-coordinate stays the same, but the *y*-coordinate becomes -2, since after a refection over the *x*-axis, the point would be located equivalently far below the *x*-axis as it is above the *x*-axis. Since the current *y*-coordinate is 2, the *y*-coordinate would become -2.

15. C: Let *r* be the number of red cans and *b* be the number of blue cans. One equation is $r + b = 10$. The total price is $16, and the prices for each can means $1r + 2b = 16$. Multiplying the first equation on both sides by -1 results in $-r - b = -10$. Add this equation to the second equation, leaving $b = 6$. So, she bought 6 *blue* cans. From the first equation, this means *r* = 4; thus, she bought 4 *red* cans.

16. B: To be directly proportional means that $y = mx$. If *x* is changed from 5 to 20, the value of *x* is multiplied by 4. Applying the same rule to the y-value, also multiply the value of *y* by 4. Therefore, *y* = 12.

17. D: Let *a* be the number of apples and *b* the number of bananas. Then, the total cost is $2a + 3b = 22$, while it also known that $a + b = 10$. Using the knowledge of systems of equations, cancel the *b* variables by multiplying the second equation by -3. This makes the equation $-3a - 3b = -30$. Adding this to the first equation, the values cancel to get $-a = -8$, which simplifies to $a = 8$.

18. D: $x \leq -5$. When solving a linear equation or inequality:

Distribution is performed if necessary:

$$-3(x + 4) \rightarrow -3x - 12 \geq x + 8$$

This means that any like terms on the same side of the equation/inequality are combined.

The equation/inequality is manipulated to get the variable on one side. In this case, subtracting *x* from both sides produces:

$$-4x - 12 \geq 8$$

The variable is isolated using inverse operations to undo addition/subtraction. Adding 12 to both sides produces $-4x \geq 20$.

The variable is isolated using inverse operations to undo multiplication/division. Remember if dividing by a negative number, the relationship of the inequality reverses, so the sign is flipped. In this case, dividing by -4 on both sides produces $x \leq -5$.

19. D: To find the average of a set of values, add the values together and then divide by the total number of values. In this case, include the unknown value of what Dwayne needs to score on his next test, in order to solve it.

$$\frac{78 + 92 + 83 + 97 + x}{5} = 90$$

Add the unknown value to the new average total, which is 5. Then multiply each side by 5 to simplify the equation, resulting in:

$$78 + 92 + 83 + 97 + x = 450$$

$$350 + x = 450$$

$$x = 100$$

Dwayne would need to get a perfect score of 100 in order to get an average of at least 90.

Test this answer by substituting back into the original formula.

$$\frac{78 + 92 + 83 + 97 + 100}{5} = 90$$

20. D: For an even number of total values, the median is calculated by finding the mean or average of the two middle values once all values have been arranged in ascending order from least to greatest. In this case, $(92 + 83) \div 2$ would equal the median 87.5, Choice *D*.

21. A: First simplify the larger fraction by separating it into two. When dividing one fraction by another, remember to invert the second fraction and multiply the two as follows:

$$\frac{5}{7} \times \frac{11}{9}$$

The resulting fraction $\frac{55}{63}$ cannot be simplified further, so this is the answer to the problem.

22. D: The slope is given by the change in *y* divided by the change in *x*. Specifically, it's:

$$slope = \frac{y_2 - y_1}{x_2 - x_1}$$

The first point is (-5,-3) and the second point is (0,-1). Work from left to right when identifying coordinates. Thus, the point on the left is point 1 (-5,-3) and the point on the right is point 2 (0,-1).

Now we need to just plug those numbers into the equation:

$$slope = \frac{-1 - (-3)}{0 - (-5)}$$

It can be simplified to:

$$slope = \frac{-1 + 3}{0 + 5}$$

$$slope = \frac{2}{5}$$

23. B: The figure is composed of three sides of a square and a semicircle. The sides of the square are simply added:

$$8 + 8 + 8 = 24\ inches$$

The circumference of a circle is found by the equation $C = 2\pi r$. The radius is 4 in, so the circumference of the circle is 25.13 in. Only half of the circle makes up the outer border of the figure (part of the perimeter) so half of 25.13 in is 12.565 in. Therefore, the total perimeter is:

$$24\ in + 12.565\ in = 36.565\ in$$

The other answer choices use the incorrect formula or fail to include all of the necessary sides.

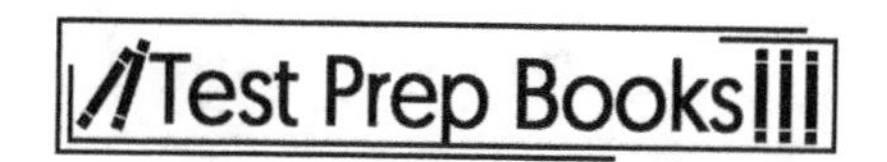

24. B: A rectangle is a specific type of parallelogram. It has 4 right angles. A square is a rhombus that has 4 right angles. Therefore, a square is always a rectangle because it has two sets of parallel lines and 4 right angles.

25. C: The number line below can be labeled with all 4 given numbers.

The number 5 is shown with the open dot. By observing the placement of the dots and their relation to the open dot at 5, the number -1 can be eliminated as an answer because it is the closest. After that, the distance to the other dots can be counted. The distance is shown below with the arcs.

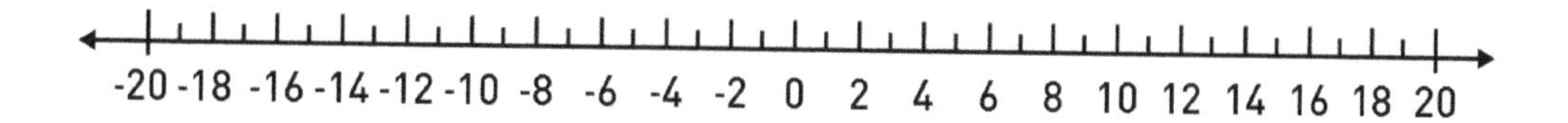

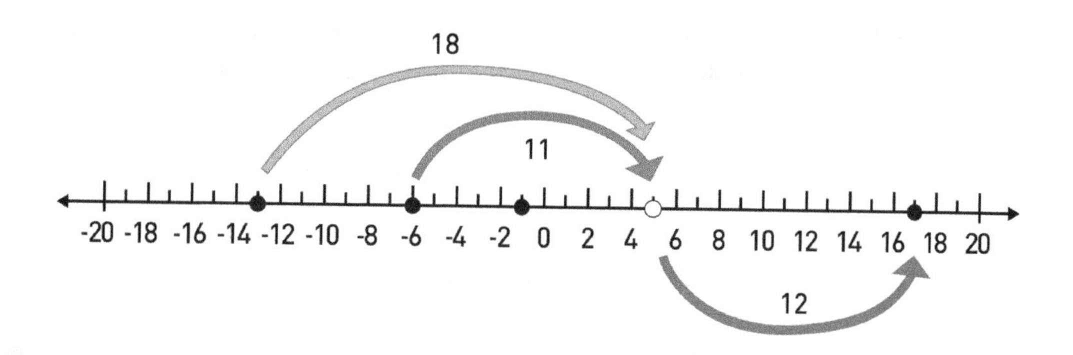

As demonstrated from the arcs showing distance on the number line, the number -13 is furthest from 5 on the number line at a distance of 18.

26. B: Start by squaring both sides to get $1 + x = 16$. Then subtract 1 from both sides to get $x = 15$.

27. A: Lining up the given scores provides the following list: 60, 75, 80, 85, and one unknown. Because the median needs to be 80, it means 80 must be the middle data point out of these five. Therefore, the unknown data point must be the fourth or fifth data point, meaning it must be greater than or equal to 80. The only answer that fails to meet this condition is 60.

28. D: The two lines are neither parallel nor perpendicular. Parallel lines will never intersect or meet. Therefore, the lines are not parallel. Perpendicular lines intersect to form a right angle (90°). Although the lines intersect, they do not form a right angle, which is usually indicated with a box at the intersection point. Therefore, the lines are not perpendicular.

29. C: The volume of a rectangular prism is $length \times width \times height$, and $2\ inches \times 4\ inches \times 6\ inches\ is\ 48\ in^3$. Choice *A* is not the correct answer because that is $2\ inches \times 6\ inches$. Choice *B* is not the correct answer because that is $4\ inches \times 6\ inches$. Choice *D* is not the correct answer because that is double of all the sides multiplied together.

30. A: The volume of a cylinder is $\pi r^2 h$, and $\pi \times 5^2 \times 10$ is $250\ \pi\ in^3$. Choice *B* is not the correct answer because that is $5^2 \times 2\pi$. Choice *C* is not the correct answer since that is $5cm \times 10\pi$. Choice *D* is not the correct answer because that is $10^2 \times 2cm$.

31. A: Let the unknown score be x. The average will be:

$$\frac{5 \times 50 + 4 \times 70 + x}{10} = \frac{530 + x}{10} = 55$$

Multiply both sides by 10 to get $530 + x = 550$, or $x = 20$.

32. B: An equilateral triangle has three sides of equal length, so if the total perimeter is 18 feet, each side must be 6 feet long. A square with sides of 6 feet will have an area of $6^2 = 36$ square feet.

33. A: Compare each numeral after the decimal point to figure out which overall number is greatest. In answers *A* (1.43785) and *C* (1.43592), both have the same tenths (4) and hundredths (3). However, the thousandths is greater in answer *A* (7), so *A* has the greatest value overall.

34. A: Operations within the parentheses must be completed first. Then, division is completed. Finally, addition is the last operation to complete. When adding decimals, digits within each place value are added together. Therefore, the expression is evaluated as:

$$(2 \times 20) \div (7 + 1) + (6 \times 0.01) + (4 \times 0.001)$$

$$40 \div 8 + 0.06 + 0.004$$

$$5 + 0.06 + 0.004 = 5.064$$

35. D: To find Denver's total snowfall, 3 must be multiplied times $27\frac{3}{4}$. In order to easily do this, the mixed number should be converted into an improper fraction.

$$27\frac{3}{4} = \frac{27 \times 4 + 3}{4} = \frac{111}{4}$$

Therefore, Denver had approximately $\frac{3 \times 111}{4} = \frac{333}{4}$ inches of snow. The improper fraction can be converted back into a mixed number through division.

$$\frac{333}{4} = 83\frac{1}{4} \text{inches}$$

36. C: $X = 150$

Set up the initial equation.

$$\frac{2X}{5} - 1 = 59$$

Add 1 to both sides.

$$\frac{2X}{5} - 1 + 1 = 59 + 1$$

Multiply both sides by $\frac{5}{2}$.

$$\frac{2X}{5} \times \frac{5}{2} = 60 \times \frac{5}{2} = 150$$

$$X = 150$$

37. C: $51.93

List the givens.

$$Tax = 6.0\% = 0.06$$

$$Sale = 50\% = 0.5$$

$$Hat = \$32.99$$

$$Jersey = \$64.99$$

Calculate the sales prices.

$$Hat\ Sale = 0.5\ (32.99) = 16.495$$

$$Jersey\ Sale = 0.5\ (64.99) = 32.495$$

Total the sales prices.

$$Hat\ sale + jersey\ sale = 16.495 + 32.495 = 48.99$$

Calculate the tax and add it to the total sales prices.

$$Total\ after\ tax = 48.99 + (48.99\ x\ 0.06\ = \$51.93$$

38. D: $0.45

List the givens.

$$Store\ coffee\ = \$1.23/lbs$$

$$Local\ roaster\ coffee\ = \$1.98/1.5\ lbs$$

Calculate the cost for 5 lbs. of store brand.

$$\frac{\$1.23}{1\ lbs} \times 5\ lbs\ = \$6.15$$

Calculate the cost for 5 lbs. of the local roaster.

$$\frac{\$1.98}{1.5\ lbs} \times 5\ lbs\ = \$6.60$$

Subtract to find the difference in price for 5 lbs.

$$\begin{array}{r} \$6.60 \\ - \underline{\quad \$6.15} \\ \$0.45 \end{array}$$

39. D: $3,325

List the givens.

$$1{,}800\ ft. = \ \$2{,}000$$

$$Cost\ after\ 1{,}800\ ft. = \ \$1.00/ft.$$

Find how many feet left after the first 1,800 ft.

$$\begin{array}{r} 3{,}125\text{ ft.} \\ - \underline{\quad 1{,}800\text{ ft.}} \\ 1{,}325\text{ ft.} \end{array}$$

Calculate the cost for the feet over 1,800 ft.

$$1{,}325\ ft. \times \frac{\$1.00}{1\ ft} = \$1{,}325$$

Add these together to find the total for the entire cost.

$$\$2{,}000\ +\ \$1{,}325\ =\ \$3{,}325$$

40. D: First, the train's journey in the real world is $3 \times 50 = 150\ miles$. On the map, 1 inch corresponds to 10 miles, so there is $\frac{150}{10} = 15$ inches on the map.

41. B: The total trip time is:

$$1 + 3.5 + 0.5 = 5\ hours$$

The total time driving is:

$$1 + 0.5 = 1.5\ hours$$

So, the fraction of time spent driving is $\frac{1.5}{5}$ or $\frac{3}{10}$. To get the percentage, convert this to a fraction out of 100. The numerator and denominator are multiplied by 10, with a result of $\frac{30}{100}$. The percentage is the numerator in a fraction out of 100, so 30%.

42. A: The value went up by:

$$\$165{,}000 - \$150{,}000 = \$15{,}000$$

Out of $150,000, this is:

$$\frac{15{,}000}{150{,}000} = \frac{1}{10}$$

Convert this to having a denominator of 100, the result is $\frac{10}{100}$ or 10%.

43. B: The journey will be $5 \times 3 = 15$ miles. A car traveling at 60 miles per hour is traveling at 1 mile per minute. The resulting equation would be:

$$\frac{15 \text{ mi}}{1 \frac{\text{mi}}{\text{min}}} = 15 \text{ min}$$

Therefore, it will take 15 minutes to make the journey.

44. B: Cone. A polygon is a closed two-dimensional figure consisting of three or more sides. A decagon is a polygon with 10 sides. A triangle is a polygon with three sides. A rhombus is a polygon with 4 sides. A cone is a three-dimensional figure and is classified as a solid.

45. C: 374.04. The formula for finding the area of a regular polygon is $A = \frac{1}{2} \times a \times P$ where *a* is the length of the apothem (from the center to any side at a right angle) and *P* is the perimeter of the figure. The apothem *a* is given as 10.39 and the perimeter can be found by multiplying the length of one side by the number of sides (since the polygon is regular):

$$P = 12 \times 6 \rightarrow P = 72$$

To find the area, substitute the values for *a* and *P* into the formula:

$$A = \frac{1}{2} \times a \times P$$

$$A = \frac{1}{2} \times (10.39) \times (72)$$

$$A = 374.04$$

46. C: Line graph. The scenario involves data consisting of two variables, month, and stock value. Box plots display data consisting of values for one variable. Therefore, a box plot is not an appropriate choice. Both line plots and circle graphs are used to display frequencies within categorical data. Neither can be used for the given scenario. Line graphs display two numerical variables on a coordinate grid and show trends among the variables.

47. D: $\frac{1}{12}$. The probability of picking the winner of the race is $\frac{1}{4}$, or $\left(\frac{number\ of\ favorable\ outcomes}{number\ of\ total\ outcomes}\right)$. Assuming the winner was picked on the first selection, three horses remain from which to choose the runner-up (these are dependent events). Therefore, the probability of picking the runner-up is $\frac{1}{3}$. To determine the probability of multiple events, the probability of each event is multiplied:

$$\frac{1}{4} \times \frac{1}{3} = \frac{1}{12}$$

Reading Comprehension

Main Idea

Recognizing the Main Idea

Typically, in narrative writing there are only a couple of ideas that the author is trying to convey to the reader. Be careful to understand the difference between a topic and a main idea. A topic might be "horses," but the main idea should be a complete sentence such as, "Racehorses run faster when they have a good relationship with the jockey." Here are some guidelines, tips, and tricks to follow that will help identify the main idea:

Identifying the Main Idea
The most important part of the text
Text title and pictures may reveal clues
Opening sentences and final sentences may reveal clues
Key vocabulary words that are repeatedly used may reveal clues

Topic versus the Main Idea

It is very important to know the difference between the topic and the main idea of the text. Even though these two are similar because they both present the central point of a text, they have distinctive differences. A **topic** is the subject of the text; it can usually be described in a one- to two-word phrase and appears in the simplest form. On the other hand, the **main idea** is more detailed and provides the author's central point of the text. It can be expressed through a complete sentence and is often found in the beginning, middle, or end of a paragraph. In most nonfiction books, the first sentence of the passage usually (but not always) states the main idea. Take a look at the passage below to review the topic versus the main idea.

> Cheetahs are one of the fastest mammals on the land, reaching up to 70 miles an hour over short distances. Even though cheetahs can run as fast as 70 miles an hour, they usually only have to run half that speed to catch up with their choice of prey. Cheetahs cannot maintain a fast pace over long periods of time because they will overheat their bodies. After a chase, cheetahs need to rest for approximately 30 minutes prior to eating or returning to any other activity.

In the example above, the topic of the passage is "Cheetahs" simply because that is the subject of the text. The main idea of the text is "Cheetahs are one of the fastest mammals on the land but can only maintain a fast pace for shorter distances." While it covers the topic, it is more detailed and refers to the text in its entirety. The text continues to provide additional details called supporting details, which will be discussed in the next section.

Theme

The **theme** of a text is the central message of the story. The theme can be about a moral or lesson that the author wants to share with the audience. Although authors do not directly state the theme of a story, it is the "big picture" that they intend readers to walk away with. For example, the fairy tale *The Boy Who Cried Wolf* features the tale of a little boy who continued to lie about seeing a wolf. When the little boy actually saw a wolf, no one believed him because of all of the previous lies. The author of this fairy tale does not directly tell readers, "Don't lie because people will question the credibility of the story." The author simply portrays the story of the little boy and presents the moral through the tale.

The theme of a text centers around varying subjects such as courage, friendship, love, bravery, facing challenges, or adversity. It often leaves readers with more questions than answers. Authors tend to insinuate certain themes in texts; however, readers are left to interpret the true meaning of the story.

Supporting Ideas

Supporting details of texts are defined as those elements of a text that help readers make sense of the main idea. They either qualitatively and/or quantitatively describe the main idea, strengthening the reader's understanding.

Supporting details answer questions like *who, what, where, when, why,* and *how*. Different types of supporting details include examples, facts and statistics, anecdotes, and sensory details.

Persuasive and informative texts often use supporting details. In persuasive texts, authors attempt to make readers agree with their points of view, and supporting details are often used as "selling points." If authors make a statement, they need to support the statement with evidence in order to adequately persuade readers. Informative texts use supporting details such as examples, facts, and details to inform readers. Take a look at the "Cheetahs" example again below to find examples of supporting details.

> Cheetahs are one of the fastest mammals on the land, reaching up to 70 miles an hour over short distances. Even though cheetahs can run as fast as 70 miles an hour, they usually only have to run half that speed to catch up with their choice of prey. Cheetahs cannot maintain a fast pace over long periods of time because they will overheat their bodies. After a chase, cheetahs need to rest for approximately 30 minutes prior to eating or returning to any other activity.

In the example above, supporting details include:

- Cheetahs reach up to 70 miles per hour over short distances.
- They usually only have to run half that speed to catch up with their prey.
- Cheetahs will overheat their bodies if they exert a high speed over longer distances.
- They need to rest for 30 minutes after a chase.

Look at the diagram below (applying the cheetah example) to help determine the hierarchy of topic, main idea, and supporting details.

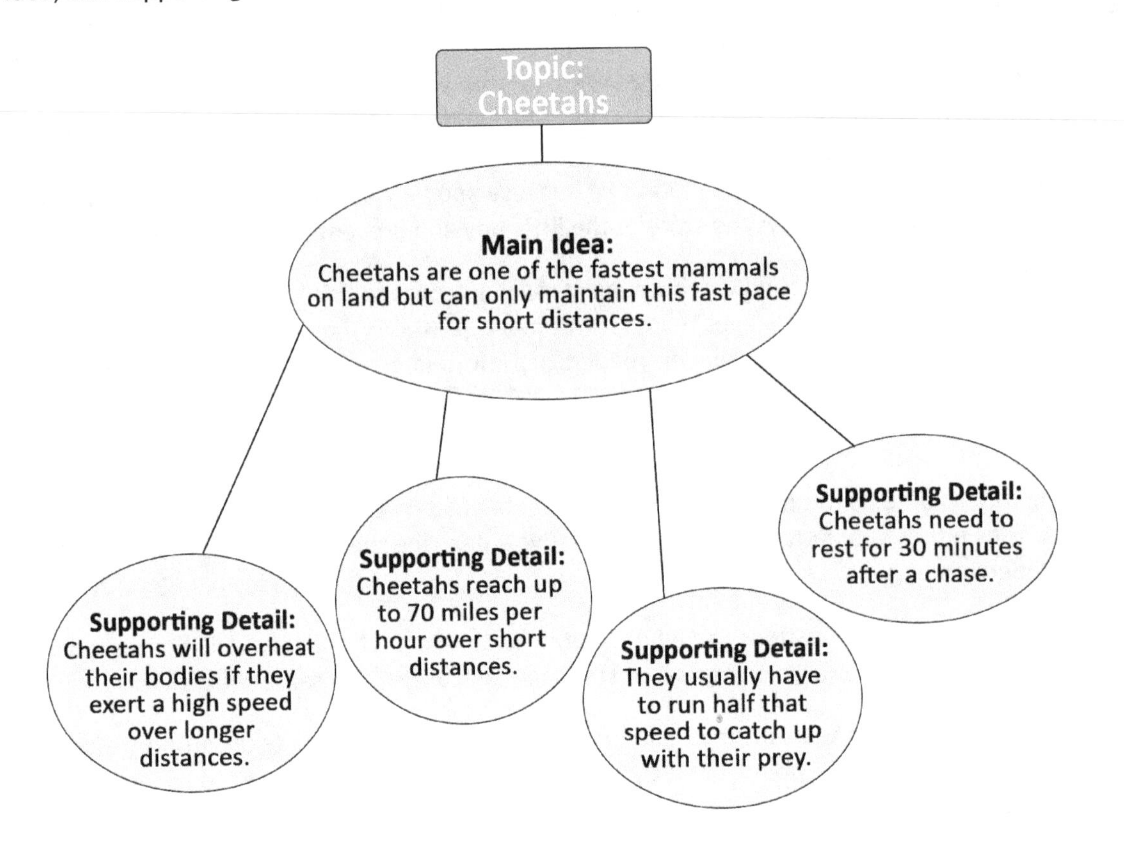

Inferences

Making Inferences

Simply put, an inference is an educated guess drawn from evidence, logic, and reasoning. The key to making inferences is identifying clues within a passage, and then using common sense to arrive at a reasonable conclusion. Consider it "reading between the lines."

One way to make an inference is to look for main topics. When doing so, pay particular attention to any titles, headlines, or opening statements made by the author. Topic sentences or repetitive ideas can be clues in gleaning inferred ideas. For example, if a passage contains the phrase *While some consider DNA testing to be infallible, it is an inherently flawed technique,* the test taker can infer the rest of the passage will contain information that points to problems with DNA testing.

The test taker may be asked to make an inference based on prior knowledge but may also be asked to make predictions based on new ideas. For example, the test taker may have no prior knowledge of DNA other than its genetic property to replicate. However, if the reader is given passages on the flaws of DNA testing with enough factual evidence, the test taker may arrive at the inferred conclusion that the author does not support the infallibility of DNA testing in all identification cases.

When making inferences, it is important to remember that the critical thinking process involved must be fluid and open to change. While a reader may infer an idea from a main topic, general statement, or other clues, he or she must be open to receiving new information within a particular passage. New ideas presented by an author may require the test taker to alter an inference. Similarly, when asked questions that require making an inference, it's important to read the entire test passage and all of the answer options. Often, a test taker will need to refine a general inference based on new ideas that may be presented within the test itself.

Inferences also refer to the ability to make logical assumptions based on clues from the text. People make inferences about the world around them on a daily basis but may not be aware of what they are doing. For example, a young boy may infer that it is likely cold outside if he wakes up and his bedroom is chilly, or the floor is cold. While being driven somewhere on the highway and a girl notices a person at the side of the road with a parked car, that girl will likely infer that the individual is having car problems and is awaiting some assistance. Both of these are example of how inferences are used every day and the same skill can be applied to different stories and texts.

In a way, making inferences is similar to detective work by collecting evidence. Sometimes clues can be found in the pictures or visual aids (like diagrams) that accompany a story or text. For example, a story may show a picture of a school in which all children are gathered in the parking lot. Upon closer examination, careful readers might spot a fire truck parked at the side of the road and may infer that the school had a fire drill or an actual fire.

Conclusions

Determining conclusions requires being an active reader, as a reader must make a prediction and analyze facts to identify a conclusion. There are a few ways to determine a logical conclusion, but careful reading is the most important. It's helpful to read a passage a few times, noting details that seem important to the text. A reader should also identify key words in a passage to determine the logical conclusion or determination that flows from the information presented.

Textual evidence within the details helps readers draw a conclusion about a passage. **Textual evidence** refers to information—facts and examples that support the main point. Textual evidence will likely come from outside sources and can be in the form of quoted or paraphrased material. In order to draw a conclusion from evidence, it's important to examine the credibility and validity of that evidence as well as how (and if) it relates to the main idea.

If an author presents a differing opinion or a **counterargument** in order to refute it, the reader should consider how and why this information is being presented. It is meant to strengthen the original argument and shouldn't be confused with the author's intended conclusion, but it should also be considered in the reader's final evaluation.

Sometimes, authors explicitly state the conclusion they want readers to understand. Alternatively, a conclusion may not be directly stated. In that case, readers must rely on the implications to form a logical conclusion:

> On the way to the bus stop, Michael realized his homework wasn't in his backpack. He ran back to the house to get it and made it back to the bus just in time.

In this example, though it's never explicitly stated, it can be inferred that Michael is a student on his way to school in the morning. When forming a conclusion from implied information, it's important to read the text carefully to find several pieces of evidence to support the conclusion.

Synthesis

Synthesis also requires a reader to make inferences while reading. Inference has been addressed earlier in this guide. Review the section and take note of the required skills in making inferences.

In order to achieve synthesis and full reading comprehension, a reader must take his or her prior knowledge, the knowledge or main ideas an author presents, and fill in the gaps to reach a logical conclusion. In a testing situation, a test taker may be asked to infer ideas from a given passage, asked to choose from a set of inferences that best express a summary of what the author hints at, or arrive at a logical conclusion based on his or her inferences. This is not an easy task, but it is approachable.

While inference requires a reader to make educated guesses based on stated information, it's important that the reader does not assume too much. It's important the reader does not insert information into a passage that's not there. It's important to make an inference based solely on the presented information and to make predictions using a logical thought process.

After reviewing the earlier section on making inferences, keep the following in mind:

- Do not jump to conclusions early in the passage. Read the full text before trying to infer meaning.
- Rely on asking questions (see the above section). What is the author stating? More important, what is the author not saying? What information is missing? What conclusions can be made about that missing information, if any?
- Make an inference then apply it back to the text passage. Does the inference make sense? Is it likely an idea with which the author would agree?
- What inferences can be made from any data presented? Are these inferences sound, logical, and do they hold water?

While this is not an exhaustive list of questions related to making inferences, it should help the reader with the skill of synthesis in achieving full reading comprehension.

Vocabulary

Deriving the Meaning of a Word or Phrase from its Context

It's common to find words that aren't familiar in writing. When you don't know a word, there are some tricks that can be used to find out its meaning. **Context clues** are words or phrases in a sentence or paragraph that provide hints about a word and what it means. For example, if an unknown word is attached to a noun with other surrounding words as clues, these can help you figure out the word's meaning. Consider the following example:

> After the treatment, Grandma's natural rosy cheeks looked *wan* and ghostlike.

The word we don't know is *wan.* The first clue to its meaning is in the phrase *After the treatment,* which tells us that something happened after a procedure (possibly medical). A second clue is the word *rosy,* which describes Grandma's natural cheek color that changed after the treatment. Finally, the word *ghostlike* infers that Grandma's cheeks now look white. By using the context clues in the sentence, we can figure out that the meaning of the word *wan* means *pale.*

Contrasts

Look for context clues that **contrast** the unknown word. When reading a sentence with a word we don't know, look for an opposite word or idea. Here's an example:

> Since Mary didn't cite her research sources, she lost significant points for *plagiarizing* the content of her report.

In this sentence, *plagiarizing* is the word we don't know. Notice that when Mary *didn't cite her research sources,* it resulted in her losing points for *plagiarizing the content of her report*. These contrasting ideas tell us that Mary did something wrong with the content. This makes sense because the definition of *plagiarizing* is "taking the work of someone else and passing it off as your own."

Contrasts often use words like *but, however, although,* or phrases like *on the other hand.* For example:

> The *gargantuan* television won't fit in my car, but it will cover the entire wall in the den.

The word we don't know is *gargantuan*. Notice that the television is too big to fit in a car, *but it will cover the entire wall in the den*. This tells us that the television is extremely large. The word *gargantuan* means *enormous*.

Synonyms

Another way to find out a word you don't know is to think of synonyms for that word. **Synonyms** are words with the same meaning. To do this, replace synonyms one at a time. Then read the sentence after each synonym to see if the meaning is clear. By replacing a word we don't know with a word we do know, it's easier to uncover its meaning. For example:

> Gary's clothes were *saturated* after he fell into the swimming pool.

In this sentence, we don't know the word *saturated*. To brainstorm synonyms for *saturated*, think about what happens to Gary's clothes after falling into the swimming pool. They'd be *soaked* or *wet*. These both turn out to be good synonyms to try. The actual meaning of *saturated* is "thoroughly soaked."

Antonyms

Sometimes sentences contain words or phrases that oppose each other. Opposite words are known as **antonyms**. An example of an antonym is *hot* and *cold*. For example:

> Although Mark seemed *tranquil*, you could tell he was actually nervous as he paced up and down the hall.

The word we don't know is *tranquil*. The sentence says that Mark was in fact not *tranquil.* He was *actually nervous*. The opposite of the word *nervous* is *calm*. *Calm* is the meaning of the word *tranquil*.

Explanations or Descriptions

Explanations or **descriptions** of other things in the sentence can also provide clues to an unfamiliar word. Take the following example:

> Golden Retrievers, Great Danes, and Pugs are the top three *breeds* competing in the dog show.

We don't know the word *breeds*. Look at the sentence for a clue. The subjects (*Golden Retrievers, Great Danes*, and *Pugs*) describe different types of dogs. This description helps uncover the meaning of the word *breeds.* The word *breeds* means "a particular type of animal."

Inferences

Inferences are clues to an unknown word that tell us its meaning. These inferences can be found within the sentence where the word appears. Or, they can be found in a sentence before the word or after the word. Look at the following example:

> The *wretched* old lady was kicked out of the restaurant. She was so mean and nasty to the waiter!

Here, we don't know the word *wretched*. The first sentence says that the *old lady was kicked out of the restaurant*, but it doesn't say why. The sentence after tells us why: *She was so mean and nasty to the waiter!* This infers that the old lady was *kicked out* because she was *so mean and nasty* or, in other words, *wretched*.

When you prepare for a vocabulary test, try reading harder materials to learn new words. If you don't know a word on the test, look for prefixes and suffixes to find out what the word means and get rid of wrong answers. If two answers both seem right, see if there are any differences between them. Then select the word that best fits. Context clues in the sentence or paragraph can also help you find the meaning of a word you don't know. By learning new words, a person can expand their knowledge. They can also improve the quality of their writing.

Understanding the Effect of Word Choice

An author's choice of words—also referred to as **diction**—helps to convey meaning in a particular way. Through diction, an author can convey a particular tone—e.g., a humorous tone, a serious tone—in order to support the thesis in a meaningful way to the reader.

Connotation and Denotation

Connotation is when an author chooses words or phrases that invoke ideas or feelings other than their literal meaning. An example of the use of connotation is the word *cheap*, which suggests something is poor in value or negatively describes a person as reluctant to spend money. When something or someone is described this way, the reader is more inclined to have a particular image or feeling. Thus, connotation can be a very effective language tool in creating emotion and swaying opinion. However, connotations are sometimes hard to pin down because varying emotions can be associated with a word. Generally, though, connotative meanings tend to be fairly consistent within a specific cultural group.

Denotation refers to words or phrases that mean exactly what they say. It is helpful when a writer wants to present hard facts or vocabulary terms with which readers may be unfamiliar. Some examples of denotation are the words *inexpensive* and *frugal*. *Inexpensive* refers to the cost of something, not its value, and *frugal* indicates that a person is conscientiously watching his or her spending. These terms do not elicit the same emotions that *cheap* does.

Authors sometimes choose to use both, but what they choose and when they use it is what critical readers need to differentiate. One method isn't inherently better than the other; however, one may create a better effect, depending upon an author's intent. If, for example, an author's purpose is to inform, to instruct, and to familiarize readers with a difficult subject, his or her use of connotation may be helpful. However, it may also undermine credibility and confuse readers. An author who wants to create a credible, scholarly effect in his or her text would most likely use denotation, which emphasizes literal, factual meaning and examples.

Technical Language

Test takers and critical readers alike should be very aware of technical language used within informational text. **Technical language** refers to terminology that is specific to a particular industry and is best understood by those specializing in that industry. This language is fairly easy to differentiate since it will most likely be unfamiliar to readers. It's critical to be able to define technical language either by the author's written definition, through the use of an included glossary—if offered—or through context clues that help readers clarify word meaning.

Organization/Logic

Types of Passages

Authors write with different purposes in mind. They use a variety of writing passages to appeal to their chosen audience. There are four types of writing passages:

- Narrative
- Expository
- Technical
- Persuasive

Each one will be described individually.

Narrative

Narrative writing tells a story or a series of events. A narrative can either be fiction or nonfiction, although in order to still be categorized as a narrative, certain elements need to be present.

A narrative must have:

- **Plot**: what happens in the story, or what is going to happen
- **Series of events**: beginning, middle, and end, but not necessarily in that order
- **Characters**: people, animals, or inanimate objects
- **Figurative language**: metaphors, similes, personification, etc.
- **Setting**: when and where the story takes place

Expository

Expository passages are informative texts usually written as memoirs or autobiographies. These nonfiction passages often use transitional words and phrases like *first, next,* and *therefore* to provide readers with a clear sense of direction of where they are within the text. Because expository passages are meant to educate readers, they often do not use flamboyant language, unless the subject area requires it.

Technical

Technical passages are written to describe how to do or make something. Technical passages are often manuals or guides written in a very organized and logical manner. The texts usually have outlines with subtitles and very little jargon. The vocabulary used in technical passages is very straightforward so as not to confuse readers. Technical texts often explore cause and effect relationships and can also include authors' purposes.

Persuasive

Persuasive passages are written with the intent to convince readers to agree with the author's viewpoint on the subject. Authors generally stick to one main point and present smaller arguments that concur with the initial central claim. The author's intent or purpose is to persuade readers by presenting them with evidence or clues (such as facts, statistics, and observations) to make them stray away from their own thoughts on the matter and agree with the author's ideas. By using evidence and clues to support their ideas, authors may or may not be strategically placing strong thoughts and emotions in their readers' minds, causing them to lean in one direction more than the other.

Readers' opinions are formed from these strong thoughts and emotions. Authors need to be wary of misleading their readers because it is not ethical, nor is it appropriate. In the same sense, readers also need to be skeptical of an author's intent if the persuasion comes from an emotional standpoint.

For example, during an election year, candidates often post slanderous advertisements about one another. These malicious ads do not state what the candidate will do for his or her country or district. They are meant to show why readers or viewers should not elect the other candidate. These ads are trying to persuade viewers' opinions of opposing candidates. On the other hand, there are other ads that show the positive impacts that candidates have made. These types of advertisements also persuade voters into liking those candidates because of the positive claims portrayed. Are these advertisements right or wrong? Are they ethical? It is up to the viewer or reader to decide.

Identifying Characteristics of Major Forms Within Each Genre

Fictional Prose

Fiction written in prose can be further broken down into **fiction genres**—types of fiction. Some of the more common genres of fiction are as follows:

- **Classical fiction**: A work of fiction considered timeless in its message or theme, remaining noteworthy and meaningful over decades or centuries—e.g., Charlotte Brontë's *Jane Eyre*, Mark Twain's *Adventures of Huckleberry Finn*

- **Fables**: Short fiction that generally features animals, fantastic creatures, or other forces within nature that assume human-like characters and has a moral lesson for the reader—e.g., *Aesop's Fables*

- **Fairy tales**: Children's stories with magical characters in imaginary, enchanted lands, usually depicting a struggle between good and evil, a sub-genre of folklore—e.g., Hans Christian Anderson's *The Little Mermaid*, *Cinderella* by the Brothers Grimm

- **Fantasy**: Fiction with magic or supernatural elements that cannot occur in the real world, sometimes involving medieval elements in language, usually includes some form of sorcery or witchcraft and sometimes set on a different world—e.g., J.R.R. Tolkien's *The Hobbit*, J.K. Rowling's *Harry Potter and the Sorcerer's Stone*, George R.R. Martin's *A Game of Thrones*

- **Folklore**: Types of fiction passed down from oral tradition, stories indigenous to a particular region or culture, with a local flavor in tone, designed to help humans cope with their condition in life and validate cultural traditions, beliefs, and customs—e.g., William Laughead's *Paul Bunyan and The Blue Ox*, the Buddhist story of "The Banyan Deer"

- **Mythology**: Closely related to folklore but more widespread, features mystical, otherworldly characters and addresses the basic question of why and how humans exist, relies heavily on allegory, and features gods or heroes captured in some sort of struggle—e.g., Greek myths, Genesis I and II in the Bible, Arthurian legends

- **Science fiction**: Fiction that uses the principle of extrapolation—loosely defined as a form of prediction—to imagine future realities and problems of the human experience—e.g., Robert Heinlein's *Stranger in a Strange Land*, Ayn Rand's *Anthem*, Isaac Asimov's *I, Robot*, Philip K. Dick's *Do Androids Dream of Electric Sheep?*

- **Short stories**: Short works of prose fiction with fully-developed themes and characters, focused on mood, generally developed with a single plot, with a short period of time for settings—e.g., Edgar Allan Poe's "Fall of the House of Usher," Shirley Jackson's "The Lottery," Isaac Bashevis Singer's "Gimpel the Fool"

Drama

Drama refers to a form of literature written for the purpose of performance for an audience. Like prose fiction, drama has several genres. The following are the most common ones:

- **Comedy**: A humorous play designed to amuse and entertain, often with an emphasis on the common person's experience, generally resolved in a positive way—e.g., Richard Sheridan's *School for Scandal*, Shakespeare's *Taming of the Shrew*, Neil Simon's *The Odd Couple*

- **History**: A play based on recorded history where the fate of a nation or kingdom is at the core of the conflict—e.g., Christopher Marlowe's *Edward II*, Shakespeare's *King Richard III*, Arthur Miller's *The Crucible*

- **Tragedy**: A serious play that often involves the downfall of the protagonist. In modern tragedies, the protagonist is not necessarily in a position of power or authority—e.g., Jean Racine's *Phèdre*, Arthur Miller's *Death of a Salesman*, John Steinbeck's *Of Mice and Men*

- **Melodrama**: A play that emphasizes heightened emotion and sensationalism, generally with stereotypical characters in exaggerated or realistic situations and with moral polarization—e.g., Jean-Jacques Rousseau's *Pygmalion*

- **Tragi-comedy**: A play that has elements of both tragedy—a character experiencing a tragic loss—and comedy—the resolution is often positive with no clear distinctive mood for either—e.g., Shakespeare's *The Merchant of Venice*, Anton Chekhov's *The Cherry Orchard*

Poetry

The genre of **poetry** refers to literary works that focus on the expression of feelings and ideas through the use of structure and linguistic rhythm to create a desired effect.

Different poetic structures and devices are used to create the various major forms of poetry. Some of the most common forms are discussed in the following chart.

Type	Poetic Structure	Example
Ballad	A poem or song passed down orally which tells a story and in English tradition usually uses an ABAB or ABCB rhyme scheme	William Butler Yeats' "The Ballad of Father O'Hart"
Epic	A long poem from ancient oral tradition which narrates the story of a legendary or heroic protagonist	Homer's *The Odyssey* Virgil's *The Aeneid*
Haiku	A Japanese poem of three unrhymed lines with five, seven, and five syllables (in English) with nature as a common subject matter	Matsuo Bashō "An old silent pond . . . A frog jumps into the pond, splash! Silence again."
Limerick	A five-line poem written in an AABBA rhyme scheme, with a witty focus	From Edward Lear's *Book of Nonsense*: "There was a Young Person of Smyrna Whose grandmother threatened to burn her . . ."
Ode	A formal lyric poem that addresses and praises a person, place, thing, or idea	Edna St. Vincent Millay's "Ode to Silence"
Sonnet	A fourteen-line poem written in iambic pentameter	Shakespeare's Sonnets 18 and 130

Literary Nonfiction

Nonfiction works are best characterized by their subject matter, which must be factual and real, describing true life experiences. There are several common types of literary non-fiction.

Biography

A **biography** is a work written about a real person (historical or currently living). It involves factual accounts of the person's life, often in a re-telling of those events based on available, researched factual information. The re-telling and dialogue, especially if related within quotes, must be accurate and reflect reliable sources. A biography reflects the time and place in which the person lived, with the goal of creating an understanding of the person and his/her human experience. Examples of well-known biographies include *The Life of Samuel Johnson* by James Boswell and *Steve Jobs* by Walter Isaacson.

Autobiography

An **autobiography** is a factual account of a person's life written by that person. It may contain some or all of the same elements as a biography, but the author is the subject matter. An autobiography will be told in first person narrative. Examples of well-known autobiographies in literature include *Night* by Elie Wiesel and *Margaret Thatcher: The Autobiography* by Margaret Thatcher.

Memoir

A **memoir** is a historical account of a person's life and experiences written by one who has personal, intimate knowledge of the information. The line between memoir, autobiography, and biography is often muddled, but generally speaking, a memoir covers a specific timeline of events as opposed to the other forms of nonfiction. A memoir is less all-encompassing. It is also less formal in tone and tends to focus on the emotional aspect of the presented timeline of events. Some examples of memoirs in literature include *Angela's Ashes* by Frank McCourt and *All Creatures Great and Small* by James Herriot.

Journalism

Some forms of **journalism** can fall into the category of literary non-fiction—e.g., travel writing, nature writing, sports writing, the interview, and sometimes, the essay. Some examples include Elizabeth Kolbert's "The Lost World, in the Annals of Extinction series for *The New Yorker* and Gary Smith's "Ali and His Entourage" for *Sports Illustrated.*

Organization of the Text

The **structure of the text** is how authors organize information in their writing. The organization of text depends on the authors' intentions and writing purposes for the text itself. Text structures can vary from paragraph to paragraph or from piece to piece, depending on the author. There are various types of patterns in which authors can organize texts, some of which include problem and solution, cause and effect, chronological order, and compare and contrast. These four text structures are described below.

Problem and Solution

One way authors can organize their text is by following a **problem and solution** pattern. This type of structure may present the problem first without offering an immediately clear solution. The problem and solution pattern may also offer the solution first and then hint at the problem throughout the text. Some texts offer multiple solutions to the same problem, which then leaves the decision as to the "best" solution to the minds of readers.

Even though the problem and solution text may seem easy to recognize, it is often confused with another organizational text pattern, cause and effect. One way of determining the difference between the two patterns is by searching for key words that indicate a problem and solution organizational text pattern is being followed, such as *propose, answer, prevention, issue, fix,* and *problematic.* Also, in problem and solution patterned texts, solutions are offered to all problems, even negative problems, unlike the cause and effect pattern.

Cause and Effect

Cause and effect is one of the more common ways that authors organize texts. In a cause and effect patterned text, the author explains what caused something to happen. For example, "It rained, so we got all wet." In this sentence, the cause is "the rain," and the effect is "we got all wet." Authors tend to use key words such as *because, as a result, due to, effected, caused, since, in order,* and *so* when writing cause and effect patterned texts.

Persuasive and expository writing models frequently use a cause and effect organizational pattern as well. In **persuasive texts**, authors try to convince readers to sway their opinions to align with the authors' thoughts. Authors may use cause and effect patterns or relationships to present supporting evidence to try to persuade the readers' judgment on a particular subject.

In **expository texts**, authors write to inform and educate readers about certain subjects. By using cause and effect patterns in expository writing, authors show relationships between events—basically, how one event may affect the other in chronological order.

Chronological Order

When using a **chronological order** organizational pattern, authors simply state information in the order in which it occurs. Nonfiction texts often include specific dates listing events in chronological order (such as timelines), whereas fiction texts may list events in order but not provide detailed dates (such as describing a daily routine: wake up, eat breakfast, get dressed, and so on). Narratives usually follow the chronological order pattern of beginning, middle, and end, with the occasional flashback in between.

Compare and Contrast

The **compare and contrast** organizational text pattern explores the differences and similarities of two or more objects. If authors describe how two or more objects are similar, they are comparing the items. If they describe how two or more objects are different, they are contrasting them. In order for texts to follow a compare and contrast organizational pattern, authors must include both similarities and differences within the text and hold each to the same guidelines. The following Venn diagram compares and contrasts oranges and apples.

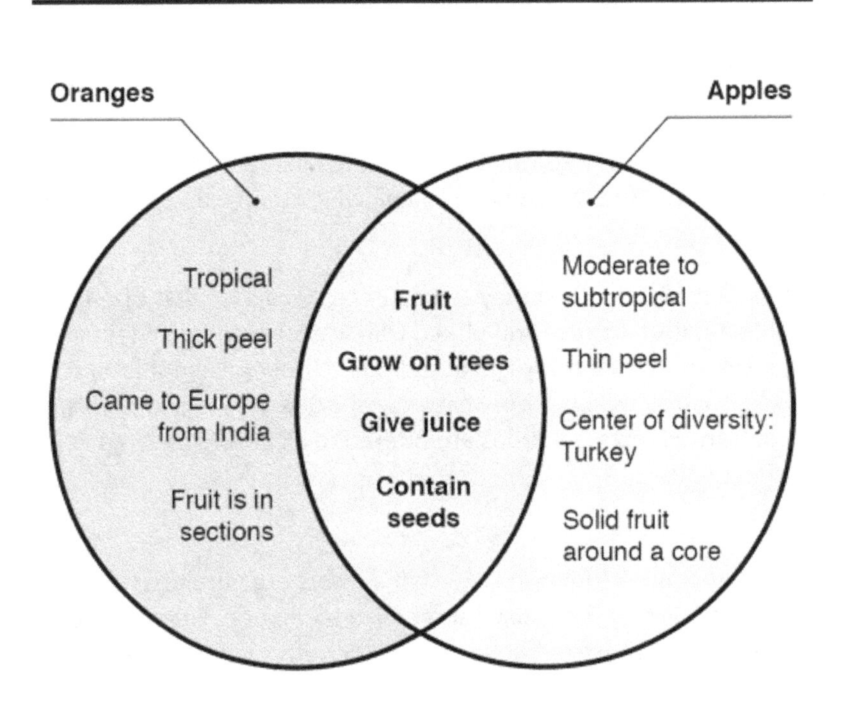

In this diagram, readers should notice how the author uses the same guidelines when comparing and contrasting oranges and apples. The origin, peel type, climate, outcomes, and classifications are all used for both fruits. The author does not use different categories for each fruit to compare or contrast the two fruits.

Even though authors may agree or disagree with each side of an argument, they should remain impartial when presenting the facts in a compare and contrast text(s). Authors should present all information using neutral language and allow readers to form their individual conclusions about the subject matter. Authors tend to use key words such as *like, unlike, different, similar, both,* and *neither* when following a compare and contrast pattern.

Overall, organization of texts helps readers better understand an author's intent and purpose.

Opinions, Facts, and Fallacies

As mentioned previously, authors write with a purpose. They adjust their writing for an intended audience. It is the readers' responsibility to comprehend the writing style or purpose of the author. When readers understand a writer's purpose, they can then form their own thoughts about the text(s) regardless of whether their thoughts are the same as or different from the author's. The following section will examine different writing tactics that authors use, such as facts versus opinions, bias and stereotypes, appealing to the readers' emotions, and fallacies.

Facts Versus Opinions

Readers need to be aware of the writer's purpose to help discern facts and opinions within texts. A **fact** is a piece of information that is true. It can either prove or disprove claims or arguments presented in texts. Facts cannot be changed or altered. For example, the statement: *Abraham Lincoln was assassinated on April 15, 1865*, is a fact. The date and related events cannot be altered.

Authors not only present facts in their writing to support or disprove their claim(s), but they may also express their opinions. Authors may use facts to support their own opinions, especially in a persuasive text; however, that does not make their opinion fact. An **opinion** is a belief or view formed about something that is not necessarily based on the truth. Opinions often express authors' personal feelings about a subject and use words like *believe, think,* or *feel.* For example, the statement: *Abraham Lincoln was the best president who has ever lived*, expresses the writer's opinion. Not all writers or readers agree or disagree with the statement. Therefore, the statement can be altered or adjusted to express opposing or supporting beliefs, such as "Abraham Lincoln was the worst president who has ever lived" or "I also think Abraham Lincoln was a great president."

When authors include facts and opinions in their writing, readers may be less influenced by the text(s). Readers need to be conscious of the distinction between facts and opinions while going through texts. Not only should the intended audience be vigilant in following authors' thoughts versus valid information, readers need to check the source of the facts presented. Facts should have reliable sources derived from credible outlets like almanacs, encyclopedias, medical journals, and so on.

Bias and Stereotypes

Not only can authors state facts or opinions in their writing, they sometimes intentionally or unintentionally show bias or portray a stereotype. A **bias** is when someone demonstrates a prejudice in favor of or against something or someone in an unfair manner. When an author is biased in his or her writing, readers should be skeptical despite the fact that the author's bias may be correct. For example, two athletes competed for the same position. One athlete is related to the coach and is a mediocre athlete, while the other player excels and deserves the position. The coach chose the less talented player who is related to him for the position. This is a biased decision because it favors someone in an unfair way.

Similar to a bias, a **stereotype** shows favoritism or opposition but toward a specific group or place. Stereotypes create an oversimplified or overgeneralized idea about a certain group, person, or place. For example,

> Women are horrible drivers.

This statement basically labels *all* women as horrible drivers. While there may be some terrible female drivers, the stereotype implies that *all* women are bad drivers when, in fact, not *all* women are. While many readers are aware of several vile ethnic, religious, and cultural stereotypes, audiences should be cautious of authors' flawed assumptions because they can be less obvious than the despicable examples that are unfortunately pervasive in society.

Fallacies

A **fallacy** is a mistaken belief or faulty reasoning, otherwise known as a **logical fallacy**. It is important for readers to recognize logical fallacies because they discredit the author's message. Readers should continuously self-question as they go through a text to identify logical fallacies. Readers cannot simply complacently take information at face value. There are six common types of logical fallacies:

- False analogy
- Circular reasoning
- False dichotomy
- Overgeneralization
- Slippery slope
- Hasty generalization

Each of the six logical fallacies are reviewed individually.

False Analogy

A **false analogy** is when the author assumes two objects or events are alike in all aspects despite the fact that they may be vastly different. Authors intend on making unfamiliar objects relatable to convince readers of something. For example, the letters *A* and *E* are both vowels; therefore, *A* = *E*. Readers cannot assume that because *A* and *E* are both vowels that they perform the same function in words or independently. If authors tell readers, *A* = *E*, then that is a false analogy. While this is a simple example, other false analogies may be less obvious.

Circular reasoning

Circular reasoning is when the reasoning is decided based upon the outcome or conclusion and then vice versa. Basically, those who use circular reasoning start out with the argument and then use false logic to try to prove it, and then, in turn, the reasoning supports the conclusion in one big circular pattern. For example, consider the two thoughts, "I don't have time to get organized" and "My disorganization is costing me time." Which is the argument? What is the conclusion? If there is not time to get organized, will more time be spent later trying to find whatever is needed? In turn, if so much time is spent looking for things, there is not time to get organized. The cycle keeps going in an endless series. One problem affects the other; therefore, there is a circular pattern of reasoning.

False Dichotomy

A **false dichotomy**, also known as a false dilemma, is when the author tries to make readers believe that there are only two options to choose from when, in fact, there are more. The author creates a false sense of the situation because he or she wants the readers to believe that his or her claim is the most logical choice. If the author does not present the readers with options, then the author is purposefully limiting what readers may believe. In turn, the author hopes that readers will believe that his or her point of view is the most sensible choice. For example, in the statement: *you either love running, or you are lazy*, the fallacy lies in the options of loving to run or being lazy. Even though both statements do not necessarily have to be true, the author tries to make one option seem more appealing than the other.

Overgeneralization

An **overgeneralization** is a logical fallacy that occurs when authors write something so extreme that it cannot be proved or disproved. Words like *all, never, most,* and *few* are commonly used when an overgeneralization is being made. For example,

> All kids are crazy when they eat sugar; therefore, my son will not have a cupcake at the birthday party.

Not *all* kids are crazy when they eat sugar, but the extreme statement can influence the readers' points of view on the subject. Readers need to be wary of overgeneralizations in texts because authors may try to sneak them in to sway the readers' opinions.

Slippery slope

A **slippery slope** is when an author implies that something will inevitably happen as a result of another action. A slippery slope may or may not be true, even though the order of events or gradations may seem logical. For example, in the children's book *If You Give a Mouse a Cookie*, the author goes off on tangents such as "If you give a mouse a cookie, he will ask for some milk. When you give him the milk, he'll probably ask you for a straw." The mouse in the story follows a series of logical events as a result of a previous action. The slippery slope continues on and on throughout the story. Even though the mouse made logical decisions, it very well could have made a different choice, changing the direction of the story.

Hasty generalization

A **hasty generalization** is when the reader comes to a conclusion without reviewing or analyzing all the evidence. It is never a good idea to make a decision without all the information, which is why hasty generalizations are considered fallacies. For example, if two friends go to a hairdresser and give the hairdresser a positive recommendation, that does not necessarily mean that a new client will have the same experience. Two referrals are not quite enough information to form an educated and well-formed conclusion.

Overall, readers should carefully review and analyze authors' arguments to identify logical fallacies and come to sensible conclusions.

Summarizing

At the end of a text or passage, it is important to summarize what the readers read. **Summarizing** is a strategy in which readers determine what is important throughout the text or passage, shorten those ideas, and rewrite or retell it in their own words. A summary should identify the main idea of the text or passage. Important details or supportive evidence should also be accurately reported in the summary. If writers provide irrelevant details in the summary, it may cloud the greater meaning of the text or passage. When summarizing, writers should not include their opinions, quotes, or what they thought the author should have said. A clear summary provides clarity of the text or passage to the readers.

The following checklist lists items that writers should include in a summary:

Summary Checklist

- Title of the story
- Someone: Who is or are the main character(s)?
- Wanted: What did the character(s) want?

- But: What was the problem?
- So: How did the character(s) solve the problem?
- Then: How did the story end? What was the resolution?

Paraphrasing

Another strategy readers can use to help them fully comprehend a text or passage is paraphrasing. **Paraphrasing** is when readers take the author's words and put them into their own words. When readers and writers paraphrase, they need to avoid copying the text—that is plagiarism. It is also important to include as many details as possible when restating the facts. Not only will this help readers and writers recall information, but by putting the information into their own words, they demonstrate if they fully comprehend the text or passage. The example below shows an original text and how to paraphrase it.

> **Original Text**: Fenway Park is home to the beloved Boston Red Sox. The stadium opened on April 20, 1912. The stadium currently seats over 37,000 fans, many of whom travel from all over the country to experience the iconic team and nostalgia of Fenway Park.
>
> **Paraphrased**: On April 20, 1912, Fenway Park opened. Home to the Boston Red Sox, the stadium now seats over 37,000 fans. Many spectators travel to watch the Red Sox and experience the spirit of Fenway Park.

Paraphrasing, summarizing, and quoting can often cross paths with one another. The chart below shows the similarities and differences between the three strategies:

Paraphrasing	Summarizing	Quoting
Uses own words	Puts main ideas into own words	Uses words that are identical to text
References original source	References original source	Requires quotation marks
Uses own sentences	Shows important ideas of source	Uses author's words and ideas

Tone/Style/Figurative Language

Tone

An author's **tone** is the use of particular words, phrases, and writing style to convey an overall meaning. Tone expresses the author's attitude towards a particular topic. For example, a historical reading passage may begin like the following:

> The presidential election of 1960 ushered in a new era, a new Camelot, a new phase of forward thinking in U.S. politics that embraced brash action, unrest, and responded with admirable leadership.

From this opening statement, a reader can draw some conclusions about the author's attitude towards President John F. Kennedy. Furthermore, the reader can make additional, educated guesses about the state of the Union during the 1960 presidential election. By close reading, the test taker can determine that the repeated use of the word *new* and words such as *admirable leadership* indicate the author's tone of admiration regarding President Kennedy's boldness. In addition, the author assesses that the era during President Kennedy's administration was problematic through the use of the words *brash action* and *unrest.* Therefore, if a test taker encountered a test question asking about the author's use of tone and their assessment of the Kennedy administration, the test taker should be able to identify an answer indicating admiration. Similarly, if asked about the state of the Union during the 1960s, a test taker should be able to correctly identify an answer indicating political unrest.

When identifying an author's tone, the following list of words may be helpful. This is not an inclusive list. Generally, parts of speech that indicate attitude will also indicate tone:

- Comical
- Angry
- Ambivalent
- Scary
- Lyrical
- Matter-of-fact
- Judgmental
- Sarcastic
- Malicious
- Objective
- Pessimistic
- Patronizing
- Gloomy
- Instructional
- Satirical
- Formal
- Casual

Style

Style can include any number of technical writing choices. A few examples of style choices include:

- Sentence Construction: When presenting facts, does the writer use shorter sentences to create a quicker sense of the supporting evidence, or do they use longer sentences to elaborate and explain the information?

- Technical Language: Does the writer use jargon to demonstrate their expertise in the subject, or do they use ordinary language to help the reader understand things in simple terms?

- Formal Language: Does the writer refrain from using contractions such as *won't* or *can't* to create a more formal tone, or do they use a colloquial, conversational style to connect to the reader?
- Formatting: Does the writer use a series of shorter paragraphs to help the reader follow a line of argument, or do they use longer paragraphs to examine an issue in great detail and demonstrate their knowledge of the topic?

Message

An author's **message** is the same as the overall meaning of a passage. It is the main idea, or the main concept the author wishes to convey. An author's message may be stated outright, or it may be implied. Regardless, the test taker will need to use careful reading skills to identify an author's message or purpose.

Often, the message of a particular passage can be determined by thinking about why the author wrote the information. Many historical passages are written to inform and to teach readers established, factual information. However, many historical works are also written to convey biased ideas to readers. Gleaning bias from an author's message in a historical passage can be difficult, especially if the reader is presented with a variety of established facts as well. Readers tend to accept historical writing as factual. This is not always the case. Any discerning reader who has tackled historical information on topics such as United States political party agendas can attest that two or more works on the same topic may have completely different messages supporting or refuting the value of the identical policies. Therefore, it is important to critically assess an author's message separate from factual information. One author, for example, may point to the rise of unorthodox political candidates in an election year based on the failures of the political party in office while another may point to the rise of the same candidates in the same election year based on the current party's successes. The historical facts of what has occurred leading up to an election year are not in refute. Labeling those facts as a failure or a success is a bias within an author's overall message, as is excluding factual information in order to further a particular point. In a standardized testing situation, a reader must be able to critically assess what the author is trying to say separate from the historical facts that surround their message.

Using the example of Lincoln's Gettysburg Address, a test question may ask the following:

> What is the message the author is trying to convey through this address?

Then they will ask the test taker to select an answer that best expresses Lincoln's message to his audience. Based on the options given, a test taker should be able to select the answer expressing the idea that Lincoln's audience should recognize the efforts of those who died in the war as a sacrifice to preserving human equality and self-government.

Effect

The **effect** an author wants to convey is when an author wants to impart a particular mood in their message. An author may want to challenge a reader's intellect, inspire imagination, or spur emotion. An author may present information to appeal to a physical, aesthetic, or transformational sense.

Take the following text as an example:

> In 1963, Martin Luther King stated, "I have a dream." The gathering at the Lincoln Memorial was the beginning of the Civil Rights movement and, with its reference to the Emancipation Proclamation, electrified those who wanted freedom and equality while rising from hatred and slavery. It was the beginning of radical change.

The test taker may be asked about the effect this statement might have on King's audience. Through careful reading of the passage, the test taker should be able to choose an answer that best identifies an effect of grabbing the audience's attention. The historical facts are in place: King made the speech in 1963 at the Lincoln Memorial, kicked off the civil rights movement, and referenced the Emancipation Proclamation. The words *electrified* and *radical change* indicate the effect the author wants the reader to understand as a result of King's speech. In this historical passage, facts are facts. However, the author's message goes beyond the facts to indicate the effect the message had on the audience and, in addition, the effect the event should have on the reader.

Figurative Language

Figurative language is a specific style of speaking or writing that uses tools for a variety of effects. It entertains readers, ignites imagination, and promotes creativity. Instead of writing in realistic terms or literal terms, figurative language plays with words and prompts readers to infer the underlying meaning. There are seven types of figurative language:

Type	Definition	Example
Personification	Giving animate qualities to an inanimate object	The tree stood tall and still, staring up at the sky.
Simile	The comparison of two unlike things using connecting words	Your eyes are as blue as the ocean.
Metaphor	The comparison of two unlike things without the use of connecting words	Kayla is a walking encyclopedia!
Hyperbole	An over-exaggeration	I could eat a million of these cookies!
Alliteration	The patterned repetition of an initial consonant sound	The bunnies are bouncing in baskets.
Onomatopoeia	Words that are formed by using the very sound associated with the word itself	"Drip, drip, drip" went the kitchen faucet.
Idioms	Common sayings that carry a lesson or meaning that must be inferred	We were sure to turn back the hands of time.

Writing Devices

Authors use a variety of writing devices throughout texts. Below is a list of some of the stylistic writing devices authors use in their writing:

- Comparison and Contrast
- Cause and Effect
- Analogy
- Point of View
- Transitional Words and Phrases

Comparison and Contrast

One writing device authors use is comparison and contrast. When authors take two objects and show how they are alike or similar, a **comparison** is being made. When authors take the same two objects and show how they differ, they are **contrasting** them. Comparison and contrast essays are most commonly written in nonfiction form. A review of the Venn diagram demonstrating common words or phrases used when comparing or contrasting objects appears below.

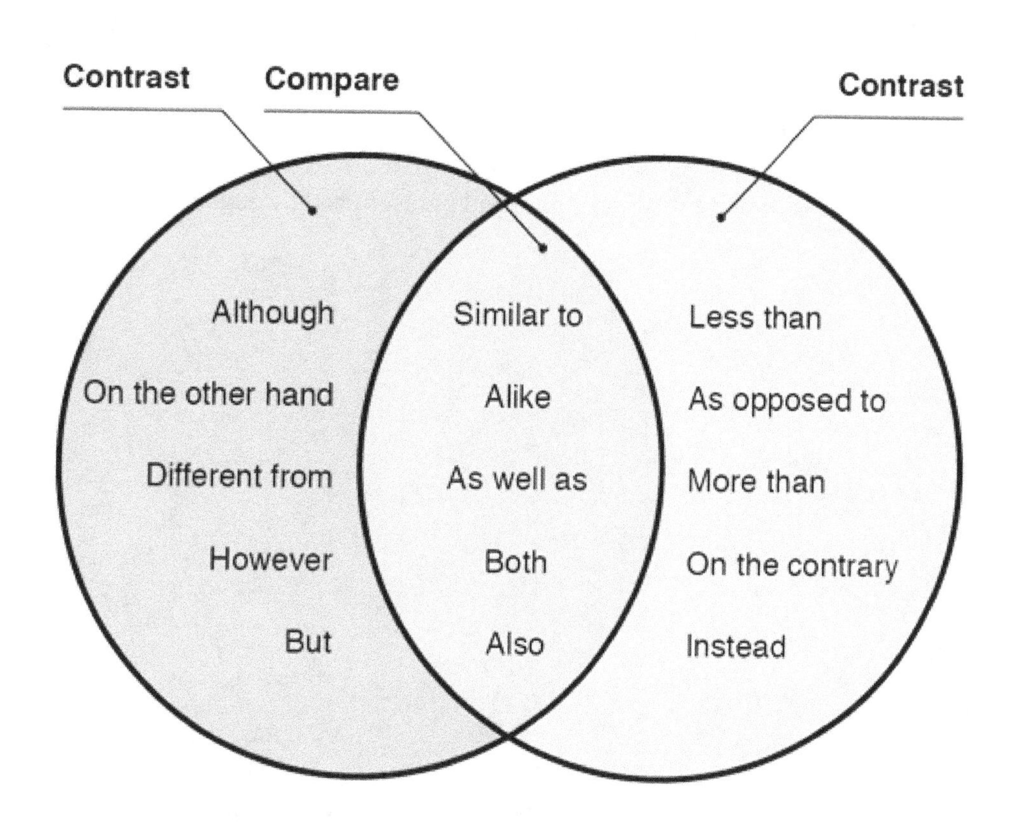

Cause and Effect

Cause and effect is one of the most common writing devices. A **cause** is why something happens, whereas an **effect** is the result that occurs because of the cause. Oftentimes, authors use key words to show cause and effect, such as *because, so, therefore, without, now, then,* and *since*. For example,

> *Because* of the sun shower, a rainbow appeared.

In this sentence, due to the sun shower (the cause), a rainbow appeared (the effect).

Analogy

An **analogy** is a comparison between two things that are quite different from one another. Authors commonly use analogies to add meaning and make ideas relatable in texts. Metaphors and similes are specific types of analogies. Metaphors compare two things that are not similar and directly connect them. Similes also compare two unlike items but connect them using the words *like* or *as*. For example,

> In the library, Alice was asked to be as quiet as a mouse.

Clearly, Alice and a mouse are very different. However, when Alice is asked to be as quiet as a mouse, readers understand that mice are small and therefore have small and soft voices—appropriate voice noise level for the library.

Point of View

Point of view is the perspective in which authors tell stories. Authors can tell stories in either the first or third person. When authors write in the first person, they are a character within a story telling about their own experiences. The pronouns *I* and *we* are used when writing in the first person. If an author writes in the third person, the narrator (the person telling the story) is telling the story from an outside perspective and is completely detached from the story. The author is not a character in the story, but rather tells about the characters' actions and dialogues. Pronouns such as *he, she, it,* and *they* are used in texts written in the third person.

Transitional Words and Phrases

There are approximately 200 transitional words and phrases that are commonly used in the English language. Below are lists of common transition words and phrases used throughout transitions.

Time

- after
- before
- during
- in the middle

Example about to be Given

- for example
- in fact
- for instance

Compare

- likewise
- also

Contrast

- however
- yet
- but

Addition

- and
- also
- furthermore
- moreover

Logical Relationships

- if
- then
- therefore
- as a result
- since

Steps in a Process

- first
- second
- last

Transitional words and phrases are important writing devices because they connect sentences and paragraphs. Transitional words and phrases present logical order to writing and provide more coherent meaning to readers.

Author's Use of Evidence to Support Claims

Authors utilize a wide range of techniques to tell a story or communicate information. Readers should be familiar with the most common of these techniques. Techniques of writing are also commonly known as rhetorical devices, and they are some of the evidence that authors use to support claims.

In nonfiction writing, authors employ argumentative techniques to present their opinion to readers in the most convincing way. Persuasive writing usually includes at least one type of appeal: an appeal to logic (**logos**), emotion (**pathos**), or credibility and trustworthiness (**ethos**). When a writer appeals to logic, they are asking readers to agree with them based on research, evidence, and an established line of reasoning. An author's argument might also appeal to readers' emotions, perhaps by including personal stories and **anecdotes** (a short narrative of a specific event). A final type of appeal, appeal to authority, asks the reader to agree with the author's argument on the basis of their expertise or credentials. Consider three different approaches to arguing the same opinion:

Logic (Logos)

Below is an example of an appeal to logic. The author uses evidence to disprove the logic of the school's rule (the rule was supposed to reduce discipline problems, but the number of problems has not been reduced; therefore, the rule is not working) and call for its repeal.

> Our school should abolish its current ban on cell phone use on campus. This rule was adopted last year as an attempt to reduce class disruptions and help students focus more on their lessons. However, since the rule was enacted, there has been no change in the number of disciplinary problems in class. Therefore, the rule is ineffective and should be done away with.

Emotion (Pathos)

An author's argument might also appeal to readers' emotions, perhaps by including personal stories and anecdotes.

The next example presents an appeal to emotion. By sharing the personal anecdote of one student and speaking about emotional topics like family relationships, the author invokes the reader's empathy in asking them to reconsider the school rule.

> Our school should abolish its current ban on cell phone use on campus. If they aren't able to use their phones during the school day, many students feel isolated from their loved ones. For example, last semester, one student's grandmother had a heart attack in the morning. However, because he couldn't use his cell phone, the student didn't know about his grandmother's accident until the end of the day—when she had already passed away, and it was too late to say goodbye. By preventing students from contacting their friends and family, our school is placing undue stress and anxiety on students.

Credibility (Ethos)

Finally, an appeal to authority includes a statement from a relevant expert. In this case, the author uses a doctor in the field of education to support the argument. All three examples begin from the same opinion—the school's phone ban needs to change—but rely on different argumentative styles to persuade the reader.

> Our school should abolish its current ban on cell phone use on campus. According to Dr. Bartholomew Everett, a leading educational expert, "Research studies show that cell phone usage has no real impact on student attentiveness. Rather, phones provide a valuable technological resource for learning. Schools need to learn how to integrate this new technology into their curriculum." Rather than banning phones altogether, our school should follow the advice of experts and allow students to use phones as part of their learning.

Practice Questions

Passage 1

[1]In recent years, some modern approaches to public transportation have made significant additions to conventional modes, such as city buses, subways, and taxis. [2]Private transportation has become convenient and affordable to anyone with access to the Internet.

[3]One of the most well-known modern transportation companies is Uber. [4]Uber employs individuals to pick up and drop off riders with the touch of a button. [5]When a rider needs to be picked up, they simply log in to the Uber app and request a car. [6]Once the app verifies the rider's location, a driver picks them up within minutes. [7]Payment for the ride is done through the app, so riders don't need to worry about carrying cash in order to catch a ride.

[8]Uber is a great resource; it is commonly used by adult commuters. [9]Smaller cities such as Deer Park, Texas, have launched their own version of Uber called DeerHaul. [10]DeerHaul operates under the same idea as Uber, only it caters to kids. [11]Because many working parents find it difficult to pick up their kids from one location and get them to another during work hours, DeerHaul is a viable alternative. [12]For example, if a child needs to be picked up from school and dropped off at Little League baseball practice but their parents can't take off work, they can arrange a ride for their child with DeerHaul. [13]All DeerHaul drivers are previous or current teachers who have passed a **battery** of fingerprint background checks. [14]Like Uber, DeerHaul rides are arranged through an Internet app, and all payments are handled electronically, so there is no money exchange between the child and the driver.

1. The main purpose of this passage is to
   a. advertise for private transportation companies such as Uber and DeerHaul.
   b. inform readers about new modes of private transportation.
   c. persuade readers to schedule a ride with DeerHaul for their children.
   d. convince readers to consider conventional methods of transportation.

2. Which statement about DeerHaul is best supported by the passage?
   a. DeerHaul is a convenient way for parents to transport their kids.
   b. DeerHaul takes no special safety measures to ensure safe rides for kids.
   c. DeerHaul is a better transportation option than Uber.
   d. Anyone can apply to be a driver for DeerHaul.

3. According to the passage, how are payments handled for private transportation?
   a. Bus or subway tickets are purchased.
   b. Riders pay the driver at the end of each ride.
   c. Tipping drivers with cash is always appreciated.
   d. Rides are paid for electronically when the ride is requested.

4. In sentence 13, the word **battery** most nearly means
   a. cell.
   b. artillery.
   c. series.
   d. violence.

5. Which sentence is the main idea of this passage?
    a. Sentence 1
    b. Sentence 2
    c. Sentence 3
    d. Sentence 8

6. Which of the following best describes the way the passage is organized?
    a. Two modes of private transportation are explained.
    b. The best method of private transportation is discussed in the passage.
    c. The analysis of one transportation company is refuted by the analysis of a second company.
    d. Conventional transportation methods are compared to more modern transportation.

Passage 2

[1]Rudolph the red-nosed reindeer hasn't always been an iconic staple in the tradition of Christmas. [2]In 1939, Montgomery Ward department stores gave Robert L. May an assignment that would unknowingly change the tradition of Christmas around the world. [3]May, a copywriter for Montgomery Ward, was asked to create a promotional Christmas coloring book. [4]The coloring book was to be a gift for children from the store Santa. [5]Montgomery Ward hoped the gift would prompt parents to bring their children to the store to visit Santa and to shop there as well.

[6]The poem, "A Visit from St. Nicholas," published in 1823, originally depicted eight flying reindeer as members of Santa's team. [7]However, in 1939, May introduced the character of Rudolph. [8]May's story depicted Rudolph as a **misfit** among the other reindeer because of his red nose. [9]Rudolph has the opportunity to prove his worth as he leads Santa's team through a blizzard on Christmas Eve. [10]For this, Rudolph becomes known as "the most famous reindeer of all." [11]At the time, Montgomery Ward officials were unsure of Rudolph's red nose because this was also known as a symptom of drunkenness. [12]However, after May and coworker Denver Gillan created a sketch of the red-nosed reindeer, officials approved the story.

[13]Montgomery Ward's promotion was a great success during the Christmas season, with 2.4 million copies of *Rudolph the Red-Nosed Reindeer* being gifted from Santas in their store. [14]In 1947, *Rudolph the Red-Nosed Reindeer* was commercially printed as a children's literature book available for purchase in bookstores.

7. What is the main idea of this passage?
    a. A Montgomery Ward advertising project made Rudolph a famous Christmas tradition.
    b. Denver Gillan first created the character of Rudolph for Montgomery Ward.
    c. Robert L. May's work helped him climb the ranks at Montgomery Ward.
    d. Without Montgomery Ward's Rudolph, Christmas would have been canceled in many countries.

8. Which statement is best supported by the passage?
    a. When first presented to the public, the character of Rudolph was associated with drunkenness.
    b. Rudolph became *the most famous reindeer of all* because he saved Christmas.
    c. Gillan is responsible for turning Way's story into a rhyming verse.
    d. The Rudolph Christmas promotional saved Montgomery Ward from having to close many stores.

9. According to the passage, *Rudolph the Red-Nosed Reindeer* was
   a. commercially printed in 1949.
   b. originally a promotional coloring book.
   c. part of an earlier poem called "A Visit from St. Nicholas."
   d. written by Denver Gillan.

10. In sentence 8, the word **misfit** most nearly means
   a. member.
   b. conformist.
   c. insider.
   d. oddball.

11. Which of the following best describes the author's tone in this passage?
   a. Humorous
   b. Apologetic
   c. Informative
   d. Sad

12. The phrase *Rudolph the red-nosed reindeer* is what type of figurative language?
   a. Simile
   b. Metaphor
   c. Alliteration
   d. Onomatopoeia

Passage 3

> [1]Water is the only substance on Earth that can occur in three states: liquid, gas, and solid. [2]The liquid state of water is the type of water you drink, cook with, or find in streams and rivers. [3]The gas form is called *water vapor,* or *steam,* and the solid, or frozen, form of water is known as *ice*.
>
> [4]Water, also known as H_2O, is made up of hydrogen and oxygen atoms. [5]These atoms join together to form water **molecules**.
>
> [6]Water molecules move at various rates of speed and distances from each other depending on the state of water. [7]When the temperature of water reaches its boiling point at 212°F, water molecules move more rapidly and spread farther apart, allowing some of them to escape into the air. [8]This turns liquid water into water vapor, or steam.
>
> [9]On the other hand, when the temperature of liquid water becomes cooler, the water molecules begin to slow down and move closer together. [10]Eventually, the water molecules stop moving and stick together to form a solid, or ice. [11]**Vice versa**, when the temperature of ice becomes warmer, water molecules begin to spread apart, causing the ice to melt and return to a liquid state.

13. According to the passage, the water molecules of boiling water
   a. move rapidly and spread farther apart.
   b. move more slowly and spread farther apart.
   c. move closer together and slow down.
   d. move rapidly and get closer together.

14. If water vapor is formed when the temperature of water reaches 212°F, what will happen to the water vapor when it rises into the air and cools?
   a. As the water vapor cools, the molecules will return to a liquid state.
   b. The water molecules will move closer together to form a solid.
   c. The water molecules will disappear into the atmosphere.
   d. Once water becomes a gas, it always remains a gas.

15. Which sentence is the main idea of this passage?
   a. Sentence 1
   b. Sentence 4
   c. Sentence 6
   d. Sentence 9

16. In sentence 4, the word **molecules** most nearly means
   a. masses.
   b. bundles.
   c. vast amounts.
   d. tiny parts.

17. According to this passage, what state of water is steam?
   a. Liquid
   b. Gas
   c. Solid
   d. Molecule

18. In sentence 11, the term **vice versa** most nearly means
   a. identically.
   b. comparably.
   c. similarly.
   d. the other way around.

Passage 4

[1]My friend Alyssa and I had spent most of the day getting ready. [2]Our moms had treated us to manicures and pedicures, followed by a fancy lunch at Burgdorf's Tea Room and then a trip to the hair salon for some curls and up-dos. [3]Alyssa and I had smiled and giggled all day in disbelief that the big day had finally arrived.

[4]At seven o'clock sharp, Ricky and Bobby came to pick us up. [5]There was an exchange of corsages and boutonnieres as our moms took an abundance of pictures to remember the night. [6]The four of us giggled and laughed as the limo driver whisked us off to the eighth-grade dance.

[7]We blared the radio and sang our favorite songs as the limo hustled into town. [8]It was all fun and games until the car suddenly swerved and we tumbled over each other in the back seat like rag dolls. [9]The tires *screeched like an alley cat* just before everything went black.

[10]It's been three months since everything went dark that night, although it feels like a lifetime. [11]The driver swerved to avoid a deer in the road. [12]He might have seen it sooner if he hadn't been texting while he was driving. [13]We never made it to the eighth-grade dance. [14]The only dancing lights I got to see that night were those of ambulances and police cars. [15]Ricky, Bobby, and Alyssa never even got to see those.

19. What is the main purpose of this passage?
    a. To persuade readers to attend their eighth-grade dance
    b. To inform readers about the availability of limo rides
    c. To illustrate the impact of texting and driving
    d. To explain how girls get ready for a dance

20. The phrase *screeched like an alley cat* in sentence 9 is an example of a
    a. metaphor.
    b. simile.
    c. alliteration.
    d. cliché.

21. This story is told in which point of view?
    a. First person
    b. Second person
    c. Third person
    d. Fourth person

22. According to the last sentence in the passage, readers can conclude that
    a. Ricky, Bobby, and Alyssa did not survive the accident.
    b. Ricky, Bobby, and Alyssa did not have to ride in an ambulance.
    c. Ricky, Bobby, and Alyssa had never planned to attend the dance.
    d. Ricky, Bobby, and Alyssa were blinded in the accident.

23. What is the overall message of this passage?
    a. Limo drivers are not required to take driving safety courses.
    b. Don't text and drive.
    c. Car accidents are common occurrences on special nights.
    d. Friends should not travel together in a limo.

24. Which statement can best be supported by this passage?
    a. The girls had been best friends since kindergarten.
    b. Ricky and Bobby came to pick up the girls in a limo.
    c. The girls' mothers were very supportive of their decision to go to the dance.
    d. The town has a great emergency response team.

Passage 5

[1]Princess Diana, born Diana Spencer on July 1, 1961, in Sandringham, England, left this world on August 31, 1997, at the age of thirty-six.

[2]Princess Diana was a graduate of Riddlesworth Hall School and West Health School. [3]After completing further schooling at Institut Alpin Videmanette in Switzerland, she worked as a teacher's assistant so she could pursue her passion for working with children.

[4]Princess Diana was known for her big heart and support of various charities. [5]Some of her most influential work included helping the homeless and children in need. [6]She also worked to provide support to people living with HIV and AIDS. [7]Princess Diana also partnered with her sons, working diligently to raise awareness in Angola about the dangers of landmines left behind after war.

[8]Princess Diana is survived by son, Prince William Arthur Philip Louis; son, Prince Henry Charles Albert David; and ex-husband, Prince Charles of Wales.

[9]Funeral services will be held September 6 at Westminster Abbey. [10]A family-only graveside ceremony will be held at the Spencer family estate, Althorp.

[11]*In lieu of* flowers, the family requests donations be made to the Diana, Princess of Wales Memorial Fund in an effort to continue the valuable charity work Princess Diana cherished.

25. What type of article is this passage?
    a. News article
    b. Preface
    c. Obituary
    d. Short story

26. In sentence 11, the phrase *in lieu of* most nearly means
    a. because of.
    b. instead of.
    c. preferably.
    d. do not send.

27. According to the information in this passage, Princess Diana spent most of her life
    a. having children.
    b. doing charity work.
    c. getting an education.
    d. funding charities.

28. After Princess Diana completed school, she
    a. moved to Sandringham, England.
    b. had children.
    c. became a teacher's assistant.
    d. did charity work.

29. According to information in the passage, what can readers infer Diana's family might do with the donations made to the Diana, Princess of Wales Memorial Fund?
    a. Pay for her funeral at the family estate.
    b. Buy clothes and school supplies for less fortunate children.
    c. Add the money to the family trust.
    d. Send her sons to college.

30. How is this passage organized?
    a. Biographical information is followed by current events.
    b. An opinion is offered about an individual.
    c. Evidence is presented to refute an individual's popularity.
    d. A thesis statement is followed by supporting paragraphs.

Passage 6

[1]Mr. Walter lived here in Vinson, Oklahoma, for his entire life. [2]He owned the little green house in town where he and his wife, Hilda, raised four boys. [3]Mr. Walter worked at the paper mill as a welder for forty-eight years.

[4]When he retired, he became the school crossing guard. [5]Every day, Mr. Walter stood on that corner and helped kids safely cross the street. [6]He knew the name of every kid who stopped at his corner, and he was always happy to see each one. [7]In the wintertime, he handed out hats and gloves to kids who'd left theirs at home. [8]Mr. Walter was always smiling as he waved to every driver that passed his corner.

[9]After he got too old to be the crossing guard, Mr. Walter bought himself a bicycle. [10]It was a three-wheeled bike, metallic red with a basket on the front. [11]At eighty years old, Mr. Walter put on his helmet and started running the roads.

[12]Mr. Walter rode that red bike around town picking up cans and trash. [13]He said it was his job to look out for his city. [14]It wasn't uncommon to see his bike parked at the McDonald's on Main. [15]He'd be standing out near the street waving at every car that passed.

[16]The whole town showed up at Mr. Walter's funeral. [17]In eighty-four years, there wasn't a day that had gone by where Mr. Walter hadn't touched one of our lives. [18]Mr. Walter had been a hero to us. [19]He'd never saved anyone from a burning house or leapt tall buildings in a single bound, but he had loved our town, and each of us in it, wholeheartedly. [20]That is what made him our hero.

31. In sentence 11, the phrase *running the roads* is an example of a(an)
    a. simile.
    b. personification.
    c. idiom.
    d. metaphor.

32. What can readers conclude from this passage?
    a. Mr. Walter retired from the paper mill and started riding his bike around town.
    b. Mr. Walter could never remember the names of the children he crossed at his corner.
    c. Mr. Walter picked up cans and trash when he worked at the paper mill.
    d. Mr. Walter touched the lives of everyone in the town of Vinson.

33. This passage can be categorized as a(n)
    a. informational text.
    b. short story.
    c. news article.
    d. persuasive letter.

34. According to the passage, it was common for Mr. Walter to
    a. give kids hats and gloves in the winter.
    b. forget to wear his safety helmet.
    c. encourage kids to pick up their trash.
    d. take his cans to the McDonald's.

35. What is the overall message of this passage?
    a. Heroes should save people from burning buildings.
    b. Sometimes it's the little things that make you a hero.
    c. Kids think all adults like Mr. Walter are heroes.
    d. All school crossing guards are special people.

36. In sentence 19, the phrase *leapt tall buildings in a single bound* is an example of a(n)
    a. hyperbole.
    b. simile.
    c. onomatopoeia.
    d. allusion.

Answer Explanations

1. B: This item asks you to determine the purpose of this passage. Choice *B* is the correct answer because the passage is written to inform readers about companies that offer more modern means of transportation. Choice *A* is incorrect because the author does not promote either company in the passage. Choice *C* is incorrect because the author does not try to persuade readers to schedule a ride for their children, and Choice *D* is incorrect because the author only makes a brief statement about conventional methods of transportation in the first sentence.

2. A: This item asks you to determine which statement can be supported based on information that is given in the passage. Choice *A* is the correct answer because sentence 12 explains how parents can use DeerHaul to help transport their children when they are unavailable. Choices *B* and *D* are incorrect because sentence 13 explains who can drive for DeerHaul and what safety precautions are taken by the company. Choice *C* is incorrect because the author offers no opinion on which company is the better option.

3. D: Your task here is to use information from the text to determine which detail is correct. Information provided in sentences 7 and 14 supports Choice *D*. Although Choice *A* may be true, buses and subways are considered public transportation. Choice *B* is incorrect, as proven in sentences 7 and 14, and Choice *C* is incorrect because tipping is not discussed in the article.

4. C: Your task here is to identify the meaning of the word *battery. Battery* most nearly means a "series," so the correct answer is Choice *C*. Although a battery can be a cell used to provide power to something, an artillery of guns, or an act of violence, none of these are the meaning that best fits this sentence. Thus, Choices *A, B,* and *D* are incorrect.

5. B: Sentence 2 is the main idea, so Choice *B* is correct. Choice *A* is incorrect because it is a supporting detail. Choices *C* and *D* are main ideas for paragraphs 2 and 3, not the entire passage.

6. A: Your task is to analyze the logical order of the information presented. Choice *A* is correct because both paragraphs 2 and 3 explain a means of private transportation. Choices *B, C,* and *D* are incorrect because the author does not offer an opinion on the best method of transportation or the best company to use.

7. A: This item asks you to determine the main idea of the passage. Choice *A* is correct because the passage discusses how a department store's advertising project made Rudolph a famous Christmas tradition. Choice *B* is mentioned in the passage as a supporting detail. Choices *C* and *D* are not mentioned in the passage.

8. B: This item asks you to determine which statement can be supported based on information given in the passage. Sentences 9 and 10 explain that Rudolph leads Santa's team through a blizzard, making him *the most famous reindeer of all*, so Choice *B* is the correct answer.

9. B: Your task here is to use information from the text to determine which detail is correct. Originally, May was asked to create a promotional coloring book, so Choice *B* is correct.

10. D: Your task here is to identify the meaning of the word *misfit*. The word *misfit* means "oddball," so Choice *D* is correct. The words *member, conformist,* and *insider* are all antonyms, or opposites, for the word *misfit,* so Choices *A, B,* and *C* are incorrect.

11. C: This item asks you to identify the author's tone in this passage. The author's tone is informative, so Choice *C* is correct. The author is not humorous, apologetic, or sad in this passage, so Choices *A, B,* and *D* are incorrect.

12. C: Your task here is to determine which type of figurative language the phrase *Rudolph the red-nosed reindeer* represents. Choice *C*, alliteration, is correct. Alliteration is the repetition of consonant letters at the beginning of one or more consecutive words. Because no comparison is made in this phrase (simile or metaphor) and the phrase is not an onomatopoeia (a word that is formed by its sound), Choices *A, B,* and *D* are incorrect.

13. A: This item asks you to identify information in the passage. Sentence 7 indicates that as the temperature of water increases, water molecules move rapidly and spread farther apart, so Choice *A* is correct.

14. A: This item asks you to draw a conclusion based on information learned in the text. Choice *A* is correct because as water vapor cools, the water molecules move closer together and move more slowly to form a liquid.

15. A: This item asks you to identify the sentence that is the main idea of this passage. Sentence 1 is the main idea because it names all three states of water. Choices *B, C,* and *D* are main ideas for support paragraphs in the passage, so they are incorrect.

16. D: Your task here is to determine the meaning of the word *molecules*. The word *molecules* means "tiny parts," as it is used in sentence 5, so Choice *D* is the correct answer. Choices *A, B,* and *C* are antonyms for the word *molecule,* so they are incorrect.

17. B: This item asks you to identify specific details in the passage. Because steam, also known as *water vapor,* is water in the form of gas, Choice *B* is correct.

18. D: This item asks you to determine the meaning of the term *vice versa*. The term *vice versa* means "the other way around," so Choice *D* is correct.

19. C: This item asks you to determine the main purpose of this passage. The author retells this sad story to illustrate the impact texting while driving can have, not just on the driver, but also on the passengers, so Choice *C* is correct.

20. B: Your task here is to identify figurative language in the passage. Choice *B* is correct because the phrase *screeched like an alley cat* is a simile. A simile is a comparison of two similar things (the sound of the tires and the screech of an alley cat) using the words *like* or *as*. Choice *A* is incorrect because a metaphor does not contain the words *like* or *as*. An alliteration is the repetition of consonant sounds at the beginning of several words, so Choice *C* is incorrect, and because this phrase is not a cliché (a phrase that is overused), Choice *D* is incorrect as well.

21. A: This item asks you to identify the story's point of view. Choice *A* is correct because the story is told from a first-person point of view. In first-person narratives, the author is a character in the story and uses words such as *I*, *me*, and *we*. Second person (Choice *B*), which is uncommon, includes words such as *you*, *your*, and *yours*. In third person (Choice *C*) narratives, the narrator is not a character in the story and uses words such as *he*, *she*, and *they*. Choice *D*, fourth person, is a made-up choice.

22. A: This task asks you to draw a conclusion based on information provided in the last sentence of the passage. The author says that Ricky, Bobby, and Alyssa never got to see the dancing lights of the ambulances and police cars. Thus, readers can conclude that they didn't survive the accident.

23. B: This task asks you to determine the overall message of this passage. Choice *B* is the correct answer since the author is trying to say don't text while driving because the results are catastrophic.

24. C: This item asks you to examine details from the story. The only choice with entirely correct information is Choice *C*. In sentence 2, the narrator explains how the girls' mothers took them to get manicures and pedicures, to a fancy lunch, and then to get their hair done before the dance. This evidence shows that both moms wanted to make the girls' day of getting ready for the dance a special one.

25. C: This item asks you to determine the type of passage. Choice *C*, an obituary, is the correct answer. Obituaries are biographies written to honor someone who has passed away.

26. B: Your task here is to determine the meaning of the phrase *in lieu of* in sentence 11. *In lieu of* means "instead of"; thus, the family is requesting that instead of sending flowers, they would prefer a donation be made to a specific charity.

27. B: This item asks you to use details in the story to determine how Princess Diana spent most of her life. Sentences 4 through 7 explain that she spent much time doing charity work, so Choice *B* is the correct answer.

28. C: This item asks you to use details in the story to determine a sequence of events. Sentence 3 explains that after school, Princess Diana became a teacher's assistant, so Choice *C* is the correct answer.

29. B: This item asks you to make an inference based on information from the passage. The passage explains that Diana did a great deal of charity work for children in need. The passage also states that the family is asking mourners to donate to the Diana, Princess of Wales Memorial Fund instead of purchasing flowers for the funeral. So, readers can infer that the family will probably use some of the money to continue Diana's work helping needy children by purchasing clothes and school supplies for them in her honor.

30. A: This item asks you to analyze the organization of the passage. Choice *A* is correct because basic information is given about Princess Diana's life, followed by current events, such as details about the funeral and where to send donations.

31. C: This item asks you to determine the meaning of a specific phrase in the passage. Choice *C* is correct because the phrase *running the roads* is an idiom, meaning "to go on adventures with friends." An idiom is a widely used expression.

32. D: The task here is to draw a conclusion based on information from the text. Throughout the passage, the author speaks of ways Mr. Walter touched the lives of everyone in the town of Vinson, so Choice *D* is correct.

33. B: This item asks you to decide how this passage would be categorized. This passage is considered a short story because it has a message (texting while driving is dangerous), a sequence of events (plot), and characters. Thus, Choice *B* is correct.

34. A: This item asks you to identify details in the passage. Choice *A* is correct because sentence 7 tells readers that Mr. Walter gave out gloves and hats during the winter to children who needed them.

35. B: This item asks you to identify the overall message in this passage. Choice *B* is correct because the author is trying to point out that heroes can be everyday people who consistently do little things in their community to impact the lives of others.

36. A: This item asks you to identify the type of figurative language used in the phrase *leapt tall buildings in a single bound.* A hyperbole is an exaggeration not meant to be taken literally, so Choice *A* is correct.

Essay

Parts of the Essay

The **introduction** has to do a few important things:

- Establish the **topic** of the essay in original wording (i.e., not just repeating the prompt)
- Clarify the significance/importance of the topic or purpose for writing. This should provide a brief overview rather than share too many details.
- Offer a **thesis statement** that identifies the writer's own viewpoint on the topic. Typically, the thesis statement is one or two brief sentences that offer a clear, concise explanation of the main point on the topic.

Body paragraphs reflect the ideas developed in the outline. Three or four points is probably sufficient for a short essay, and they should include the following:

- A **topic sentence** that identifies the sub-point (e.g., a reason why, a way how, a cause or effect)
- A detailed **explanation** of each sub-point, explaining why the writer thinks this point is valid
- **Illustrative examples**, such as personal examples or real-world examples, that support and validate the point (i.e., "prove" the point)
- A **concluding sentence** that connects the examples, reasoning, and analysis to the point being made

The **conclusion,** or final paragraph, should be brief and should reiterate the focus, clarifying why the discussion is significant or important. It is important to avoid adding specific details or new ideas to this paragraph. The purpose of the conclusion is to sum up what has been said to bring the discussion to a close.

The Short Overview

The essay may seem challenging, but following these steps can help writers focus:

- Take one to two minutes to think about the topic.
- Generate some ideas through brainstorming (three to four minutes).
- Organize ideas into a brief outline, selecting just three to four main points to cover in the essay
- Develop essay in parts:
    - Introduction paragraph, with intro to topic and main points
    - Viewpoint on the subject at the end of the introduction
    - Body paragraphs, based on outline, each should make a main point, explain the viewpoint, and use examples to support the point
    - Brief conclusion highlighting the main points and closing

- Read over the essay (last five minutes).
- Look for any obvious errors, making sure that the writing makes sense.

Writing an essay can be overwhelming, and performance panic is a natural response. The outline serves as a basis for the writing and help writers keep focused. Getting stuck can also happen, and it's helpful to remember that brainstorming can be done at any time during the writing process. Following the steps of the writing process is the best defense against writer's block.

Timed essays can be particularly stressful, but assessors are trained to recognize the necessary planning and thinking for these timed efforts. Using the plan above and sticking to it helps with time management. Timing each part of the process helps writers stay on track. Sometimes writers try to cover too much in their essays. If time seems to be running out, this is an opportunity to determine whether all of the ideas in the outline are necessary. Three body paragraphs are sufficient, and more than that is probably too much to cover in a short essay.

More isn't always *better* in writing. A strong essay will be clear and concise. It will avoid unnecessary or repetitive details. It is better to have a concise, five-paragraph essay that makes a clear point, than a ten-paragraph essay that doesn't. The goal is to write one to two pages of quality writing. Paragraphs should also reflect balance; if the introduction goes to the bottom of the first page, the writing may be going off-track or may be repetitive. It's best to fall into the one to two-page range, but a complete, well-developed essay is the ultimate goal.

Applying Basic Knowledge of the Elements of the Writing Process

Practice Makes Prepared Writers

Like any other useful skill, writing only improves with practice. While writing may come more easily to some than others, it is still a skill to be honed and improved. Regardless of a person's natural abilities, there is always room for growth in writing. Practicing the basic skills of writing can aid in preparations for the ISEE.

One way to build vocabulary and enhance exposure to the written word is through reading. This can be through reading books, but reading of any materials such as newspapers, magazines, and even social media count towards practice with the written word. This also helps to enhance critical reading and thinking skills, through analysis of the ideas and concepts read. Think of each new reading experience as a chance to sharpen these skills.

Planning

Brainstorming

One of the most important steps in writing an essay is prewriting. Before drafting an essay, it's helpful to think about the topic for a moment or two, in order to gain a more solid understanding of what the task is. Then, spending about five minutes jotting down the immediate ideas that could work for the essay is recommended. Brainstorming is a way to get some words on the page and offer a reference for ideas when drafting. Scratch paper is provided for writers to use any prewriting techniques such as webbing, free writing, or listing. The goal is to get ideas out of the mind and onto the page.

In the planning stage, it's important to consider all aspects of the topic, including different viewpoints on the subject. There are more than two ways to look at a topic, and a strong argument considers those opposing viewpoints. Considering opposing viewpoints can help writers present a fair, balanced, and informed essay that shows consideration for all readers. This approach can also strengthen an argument by recognizing and potentially refuting the opposing viewpoint(s).

Drawing from personal experience may help to support ideas. For example, if the goal for writing is a personal narrative, then the story should be from the writer's own life. Many writers find it helpful to draw from personal experience, even in an essay that is not strictly narrative. Personal anecdotes or short stories can help to illustrate a point in other types of essays as well.

Once the ideas are on the page, it's time to turn them into a solid plan for the essay. The best ideas from the brainstorming results can then be developed into a more formal outline.

Outlining

An **outline** is a system used to organize writing. When reading texts, outlining is important because it helps readers organize important information in a logical pattern using Roman numerals. Usually, outlines start out with the main idea(s) and then branch out into subgroups or subsidiary thoughts or subjects. The outline should be methodical, with at least two main points followed each by at least two subpoints. Outlines provide a visual tool for readers to reflect on how events, characters, settings, or other key parts of the text or passage relate to one another. They can also lead readers to a stronger conclusion. The sample below demonstrates what a general outline looks like.

I. Main Topic 1
   a. Subtopic 1
   b. Subtopic 2
      1. Detail 1
      2. Detail 1
II. Main Topic 2
   a. Subtopic 1
   b. Subtopic 2
      1. Detail 1
      2. Detail 2

Free Writing

Like brainstorming, **free writing** is another prewriting activity to help the writer generate ideas. This method involves setting a timer for 2 or 3 minutes and writing down all ideas that come to mind about the topic using complete sentences. Once time is up, review the sentences to see what observations have been made and how these ideas might translate into a more coherent direction for the topic. Even if sentences lack sense as a whole, this is an excellent way to get ideas onto the page in the very beginning stages of writing. Using complete sentences can make this a bit more challenging than brainstorming, but overall it is a worthwhile exercise, as it may force the writer to come up with more complete thoughts about the topic.

Writing

Now it comes time to actually write your essay. Follow the outline you developed in the brainstorming process and try to incorporate the sentences you wrote in the free writing exercise.

Basing the essay on the outline aids in both organization and coherence. The goal is to ensure that there is enough time to develop each sub-point in the essay, roughly spending an equal amount of time on each idea. Keeping an eye on the time will help. If there are fifteen minutes left to draft the essay, then it makes sense to spend about 5 minutes on each of the ideas. Staying on task is critical to success and timing out the parts of the essay can help writers avoid feeling overwhelmed.

Remember that your work here does not have to be perfect. This process is often referred to as **drafting** because you're just creating a rough draft of your work.

Don't get bogged down on the small details. For instance, if you're not sure whether or not a word should be capitalized, mark it somehow and look up the capitalization rule while in the revision process if not in a testing situation. The same goes for referencing sources. That should not be focused on until after the writing process.

Referencing Sources

Anytime you quote or paraphrase another piece of writing you will need to include a citation. A **citation** is a short description of the work that your quote or information came from. The manual of style your teacher wants you to follow will dictate exactly how to format that citation. For example, this is how you would cite a book according to the APA manual of style:

- *Format*: Last name, First initial, Middle initial. (Year Published) *Book Title.* City, State: Publisher.
- *Example*: Sampson, M. R. (1989). *Diaries from an alien invasion*. Springfield, IL: Campbell Press.

Revising

Revising and proofreading offers an opportunity for writers to polish things up. Putting one's self in the reader's shoes and focusing on what the essay actually says helps writers identify problems—it's a movement from the mindset of writer to the mindset of editor. The goal is to have a clean, clear copy of the essay.

During the essay portion of a test, leaving a few minutes at the end to revise and proofread offers an opportunity for writers to polish things up. Putting one's self in the reader's shoes and focusing on what the essay actually says helps writers identify problems—it's a movement from the mindset of writer to the mindset of editor. The goal is to have a clean, clear copy of the essay. The following areas should be considered when proofreading:

- Sentence fragments
- Awkward sentence structure
- Run-on sentences
- Incorrect word choice
- Grammatical agreement errors
- Spelling errors
- Punctuation errors
- Capitalization errors

Recursive Writing Process

While the writing process may have specific steps, the good news is that the process is recursive, meaning the steps need not be completed in a particular order. Many writers find that they complete steps at the same time such as drafting and revising, where the writing and rearranging of ideas occur simultaneously or in very close order. Similarly, a writer may find that a particular section of a draft needs more development and will go back to the prewriting stage to generate new ideas. The steps can be repeated at any time, and the more these steps of the recursive writing process are employed, the better the final product will be.

Developing a Well-Organized Paragraph

Forming Paragraphs

A good **paragraph** should have the following characteristics:

- Be logical with organized sentences
- Have a unified purpose within itself
- Use sentences as building blocks
- Be a distinct section of a piece of writing
- Present a single theme introduced by a topic sentence
- Maintain a consistent flow through subsequent, relevant, well-placed sentences
- Tell a story of its own or have its own purpose, yet connect with what is written before and after
- Enlighten, entertain, and/or inform

Though certainly not set in stone, the length should be a consideration for the reader's sake, not merely for the sake of the topic. When paragraphs are especially short, the reader might experience an irregular, uneven effect; when they're much longer than 250 words, the reader's attention span, and probably their retention, is challenged. While a paragraph can technically be a sentence long, a good rule of thumb is for paragraphs to be at least three sentences long and no more than ten sentence long. An optimal word length is 100 to 250 words.

Coherent Paragraphs

Coherence is simply defined as the quality of being logical and consistent. In order to have coherent paragraphs, therefore, authors must be logical and consistent in their writing, whatever the document might be. Two words are helpful to understanding coherence: flow and relationship. Earlier, transitions were referred to as being the "glue" to put organized thoughts together. Now, let's look at the topic sentence from which flow and relationship originate.

The **topic sentence**, usually the first in a paragraph, holds the essential features that will be brought forth in the paragraph. It is also here that authors either grab or lose readers. It may be the only writing that a reader encounters from that writer, so it is a good idea to summarize and represent ideas accurately.

The coherent paragraph has a logical order. It utilizes transitional words and phrases, parallel sentence structure, clear pronoun references, and reasonable repetition of key words and phrases. Use common sense for repetition. Consider synonyms for variety. Be consistent in verb tense whenever possible.

When writers have accomplished their paragraph's purpose, they prepare it to receive the next paragraph. While writing, read the paragraph over, edit, examine, evaluate, and make changes accordingly. Possibly, a paragraph has gone on too long. If that occurs, it needs to be broken up into other paragraphs, or the length should be reduced. If a paragraph didn't fully accomplish its purpose, consider revising it.

Main Point of a Paragraph

What is the main point of a paragraph? It is *the* point all of the other important and lesser important points should lead up to, and it should be summed up in the topic sentence.

Sometimes there is a fine line between a paragraph's topic sentence and its main point. In fact, they actually might be one and the same. Often, though, they are two separate, but closely related, aspects of the same paragraph.

Depending upon the writer's purpose, the topic sentence or the paragraph's main point might not be fully revealed until the paragraph's conclusion.

Sometimes, while developing paragraphs, authors deviate from the main point, which means they have to delete and rework their materials to stay on point.

Examining Paragraphs

While authors write, thoughts coalesce to form words on "paper." Authors strategically place those thoughts in sentences to give them "voice" in an orderly manner, and then they manipulate them into cohesive sentences for cohesion to express ideas. Like a hunk of modeling clay, sentences can be worked and reworked until they cooperate and say what was originally intended.

Before calling a paragraph complete, identify its main point, making sure that related sentences stay on point. Pose questions such as, "Did I sufficiently develop the main point? Did I say it succinctly enough? Did I give it time to develop? *Is* it developed?"

Let's examine the following two paragraphs, each an example of a movie review. Read them and form a critique.

> Example 1: *Eddie the Eagle* is a movie about a struggling athlete. Eddie was crippled at birth. He had a lot of therapy and he had a dream. Eddie trained himself for the Olympics. He went far away to learn how to ski jump. It was hard for him, but he persevered. He got a coach and kept trying. He qualified for the Olympics. He was the only one from Britain who could jump. When he succeeded, they named him, "Eddie the Eagle."

> Example 2: The last movie I saw in the theater was *Eddie the Eagle,* a story of extraordinary perseverance inspired by real life events. Eddie was born in England with a birth defect that he slowly but surely overcame, but not without trial and error (not the least of which was his father's perpetual *dis*couragement). In fact, the old man did everything to get him to give up, but Eddie was dogged beyond anyone in the neighborhood; in fact, maybe beyond anyone in the whole town or even the whole world! Eddie, simply, did not know to quit. As he grew up, so did his dream; a strange one, indeed, for someone so unaccomplished: to compete in the Winter Olympics as a ski jumper (which he knew absolutely nothing about). Eddie didn't just keep on dreaming about it. He actually went to Germany and *worked* at it, facing unbelievable odds, defeats, and put-downs by Dad and the other Men in Charge, aka the Olympic decision-makers. Did that stop him? No way! Eddie got a coach and persevered. Then, when he failed, he persevered some more, again and again. You should be able to open up a dictionary, look at the word "persevere," and see a picture of Eddie the Eagle because, when everybody told him he couldn't, he did. The result? He is forever dubbed, "Eddie the Eagle."

Both reviews tell something about the movie *Eddie the Eagle*. Does one motivate the reader to want to see the movie more than the other? Does one just provide a few facts while the other paints a virtual picture of the movie? Does one give a carrot and the other a rib eye steak, mashed potatoes, and chocolate silk pie?

Paragraphs sometimes only give facts. Sometimes that's appropriate and all that is needed. Sometimes, though, writers want to use the blank documents on their computer screens to paint a picture. Writers must "see" the painting come to life. To do so, pick a familiar topic, write a simple sentence, and add to it. Pretend, for instance, there's a lovely view. What does one see? Is it a lake? Try again—picture it as though it were the sea! Visualize a big ship sailing out there. Is it sailing away or approaching? Who is on it? Is it dangerous? Is it night and are there crazy pirates on board? Uh-oh! Did one just jump ship and start swimming toward shore?

Distinguishing Between Formal and Informal Language

It can be helpful to distinguish whether a writer or speaker is using formal or informal language because it can give the reader or listener clues to whether the text is informative, nonfiction, argumentative, or the intended tone or audience. Formal and informal language in written or verbal communication serve different purposes and are often intended for different audiences. Consequently, their tone, word choices, and grammatical structures vary. These differences can be used to identify which form of language is used in a given piece and to determine which type of language should be used for a certain context. Understanding the differences between formal and informal language will also allow a writer or speaker to implement the most appropriate and effective style for a given situation.

Formal language is less personal and more informative and pragmatic than informal language. It is more "buttoned-up" and business-like, adhering to proper grammatical rules. It is used in professional or academic contexts, to convey respect or authority. For example, one would use formal language to write an informative or argumentative essay for school and to address a superior or esteemed professional like a potential employer, a professor, or a manager. Formal language avoids contractions, slang, colloquialisms, and first-person pronouns. **Slang** refers to non-standard expressions that are not used in elevated speech and writing. Slang creates linguistic in-groups and out-groups of people, those who can understand the slang terms and those who can't. Slang is often tied to a specific time period. For example, "groovy" and "far out" are connected to the 1970s, and "as if!" and "4-1-1-" are connected to the 1990s. **Colloquial language** is language that is used conversationally or familiarly—e.g., "What's up?"—in contrast to formal, professional, or academic language—"How are you this evening?" Formal language uses sentences that are usually more complex and often in passive voice. Punctuation can differ as well. For example, **exclamations point (!)** are used to show strong emotion or can be used as an interjection but should be used sparingly in formal writing situations.

Informal language is often used when communicating with family members, friends, peers, and those known more personally. It is more casual, spontaneous, and forgiving in its conformity to grammatical rules and conventions. Informal language is used for personal emails, some light fiction stories, and some correspondence between coworkers or other familial relationships. The tone is more relaxed and slang, contractions, clichés, and the first and second person may be used in writing. The imperative voice may be used as well.

As a review, the perspectives from which something may be written or conveyed are detailed below:

- First-person point of view: The story is told from the writer's perspective. In fiction, this would mean that the main character is also the narrator. First-person point of view is easily recognized by the use of personal pronouns such as *I, me, we, us, our, my*, and *myself.*
- Second-person point of view: This point of view isn't commonly used in fiction or nonfiction writing because it directly addresses the reader using the pronouns *you, your*, and *yourself.*

Second-person perspective is more appropriate in direct communication, such as business letters or emails.

- Third-person point of view: In a more formal essay, this would be an appropriate perspective because the focus should be on the subject matter, not the writer or the reader. Third-person point of view is recognized by the use of the pronouns *he, she, they*, and *it*. In fiction writing, third-person point of view has a few variations.
    - Third-person limited point of view refers to a story told by a narrator who has access to the thoughts and feelings of just one character.
    - In third-person omniscient point of view, the narrator has access to the thoughts and feelings of all the characters.
    - In third-person objective point of view, the narrator is like a fly on the wall and can see and hear what the characters do and say but does not have access to their thoughts and feelings.

Tips for the Essay

- Don't panic! This section isn't scored. It's just a great way to show teachers how smart you are and how well you can tell a story and write. You can do it!
- Use your time well. Fifteen minutes is quick! It is shorter than almost every TV show. Don't spend too much time doing any one thing. Try to brainstorm briefly and then get writing. Leave a few minutes to read it over and correct any spelling mistakes or confusing parts.
- Be yourself! You are smart and interesting, and teachers want to get to know you and your unique ideas. Don't feel pressured to use big vocabulary words if you aren't positive what they mean. You will be more understandable if you use the right word, not the fanciest word.

Practice Essay

You have 30 minutes to plan and write your essay. Do not worry too much about length; it's most important to focus on the content and quality of your writing. Choose one of the following three questions to answer in a coherent essay:

- What would you like to pursue as a career and why?
- Where in the world would you most like to have a pen pal and why?
- If you were asked to mentor younger children in a skill or hobby you enjoy, what would you share with them? How has that activity influenced your life?

Dear Customer,

We would like to start by thanking you for purchasing this study guide for the ISEE Middle Level exam. We hope that we exceeded your expectations.

Our goal in creating this study guide was to cover all of the topics that you will see on the test. We also strove to make our practice questions as similar as possible to what you will encounter on test day. With that being said, if you found something that you feel was not up to your standards, please send us an email and let us know.

We would also like to let you know about another book in our catalog that may interest you.

PSAT 8/9:

This can be found on Amazon: amazon.com/dp/1628456612

SSAT Middle:

This can be found on Amazon: amazon.com/dp/1628458968

We have study guides in a wide variety of fields. If the one you are looking for isn't listed above, then try searching for it on Amazon or send us an email.

Thanks Again and Happy Testing!
Product Development Team
info@studyguideteam.com

FREE Test Taking Tips DVD Offer

To help us better serve you, we have developed a Test Taking Tips DVD that we would like to give you for FREE. **This DVD covers world-class test taking tips that you can use to be even more successful when you are taking your test.**

All that we ask is that you email us your feedback about your study guide. Please let us know what you thought about it – whether that is good, bad or indifferent.

To get your **FREE Test Taking Tips DVD**, email freedvd@studyguideteam.com with "FREE DVD" in the subject line and the following information in the body of the email:

a. The title of your study guide.

b. Your product rating on a scale of 1-5, with 5 being the highest rating.

c. Your feedback about the study guide. What did you think of it?

d. Your full name and shipping address to send your free DVD.

If you have any questions or concerns, please don't hesitate to contact us at freedvd@studyguideteam.com.

Thanks again!

Printed in the USA
CPSIA information can be obtained
at www.ICGtesting.com
CBHW081527220824
13574CB00012B/659

9 781628 45739